Smart Farming - Integrating Conservation Agriculture, Information Technology, and Advanced Techniques for Sustainable Crop Production

Edited by Subhan Danish, Hakoomat Ali and Rahul Datta

Published in London, United Kingdom

Smart Farming - Integrating Conservation Agriculture, Information Technology, and Advanced Techniques for Sustainable Crop Production
http://dx.doi.org/10.5772/intechopen.102141
Edited by Subhan Danish, Hakoomat Ali and Rahul Datta

Contributors
Pomi Shahbaz, Shamsheer ul Haq, Ismet Boz, Alexy Márta, András Jung, Bálint Molnár, Raja Shankar, Panamanna Mahadevan Govindakrishnan, Shashi Rawat, Joseph Sherly, Mohammed Nagib Abd El-Ghany Hasaneen, Subhan Danish, Hakoomat Ali, Rahul Datta

First published in London, United Kingdom, 2023 by IntechOpen
IntechOpen is the global imprint of INTECHOPEN LIMITED, registered in England and Wales, registration number: 11086078, 5 Princes Gate Court, London, SW7 2QJ, United Kingdom

British Library Cataloguing-in-Publication Data
A catalogue record for this book is available from the British Library

Additional hard and PDF copies can be obtained from orders@intechopen.com

Smart Farming - Integrating Conservation Agriculture, Information Technology, and Advanced Techniques for Sustainable Crop Production
Edited by Subhan Danish, Hakoomat Ali and Rahul Datta
p. cm.
Print ISBN 978-1-80356-689-4
Online ISBN 978-1-80356-690-0
eBook (PDF) ISBN 978-1-80356-691-7

Meet the editors

Dr. Subhan Danish is a guest lecturer at Bahauddin Zakariya University, Pakistan. He has a bachelor's degree in agriculture, a master's degree and a Ph.D. in agriculture (soil science). Dr. Danish teaches at undergraduate and graduate levels, focusing on soil science and plant nutrition. His current research interests include soil microbial and nutrient dynamics, their regulation pattern with non-coding and their role in plant growth, abiotic stress alleviation, nanoparticles in agriculture, and climate-smart agriculture with artificial intelligence. His papers have been published in a variety of peer-reviewed journals, including *Scientific Report, Journal of Environmental Management, Science of the Total Environment, Cleaner Production*, and Chemosphere. His work has been cited over 3980 times by researchers from all over the world and has an H-index of 32 and i10-index of 98. He is a reviewer for a number of international journals: *Archives of Agronomy and Soil Science, Environmental Science and Pollution Research, Journal of Environmental Management, Remote Sensing, Scientific Report, Environmental Technology, and Innovation and Frontiers in Plant Science.*

Dr. Hakoomat Ali is a Professor of Agronomy and is currently the Dean of the Faculty of Agricultural Sciences and Technology, Bahauddin Zakariya University, Pakistan, where he established the Advanced Crop Nutrition Laboratory. He obtained his Ph.D. in crop nutrition and physiology from the University of Wales, Aberystwyth, UK, having received a Merit Scholarship for a Ph.D. undertaken abroad from Pakistan's Ministry of Education in 1994. He has supervised over 100 MPhil and Ph.D. students and received the National Best Teacher Award for the year 2014 from the Islamabad Higher Education Commission. His current research focuses on nutrient management of arable crops and he is working to devise production technologies for different crops against the background of worldwide climate change. He has won many valuable research projects from national and international funding agencies. He has published about 127 articles in peer-reviewed journals and eight book chapters. According to Scopus®, his publications have received roughly 2200+ citations with an H-index of 31 and i10-index of 52. He is an editor and reviewer for more than 100 peer-reviewed international journals as well as editor-in-chief of the *Journal of Arable Crops and Marketing.* He has been honored by different authorities for his outstanding performance in research and education and received a Research Productivity Award from the Pakistan Higher Education Commission in 2011 and 2014.

Dr. Rahul Datta is an assistant professor at Mendel University, Czech Republic, where he obtained his MSc and Ph.D. He is a soil microbiologist and enzymologist. His research is focused on sustainable agriculture and food security, with specific reference to the application of biochar, soil molecular and microbiological techniques to improve soil quality and crop productivity. He has published about 75 research articles and review papers and two books with Springer and IntechOpen, and has reviewed more than 300 papers in peer-reviewed journals. At present, he is an Honorary International Advisor at the Asian PGPR Society.

Contents

Preface **IX**

Section 1
Introduction **1**

Chapter 1 **3**
Introductory Chapter: Smart Farming
by Subhan Danish, Hakoomat Ali and Rahul Datta

Section 2
Sustainable Farming **9**

Chapter 2 **11**
Adoption of Conservation Agriculture as a Driver of Sustainable Farming: Opportunities, Constraints, and Policy Issues
by Pomi Shahbaz, Shamsheer ul Haq and Ismet Boz

Section 3
Information Technology Use in Smart Farming **29**

Chapter 3 **31**
Information Technology Drivers in Smart Farming Management Systems
by Alexy Márta, András Jung and Bálint Molnár

Section 4
Smart Farming Approach **57**

Chapter 4 **59**
Perspective Chapter: Physiological Breeding Approach for Sustainable Smart Farming
by Raja Shankar, Panamanna Mahadevan Govindakrishnan, Shashi Rawat and Joseph Sherly

Chapter 5 **87**
Perspective Chapter: Recent Advances in Nanotechnology, Nanomaterials, Nanofertilizers and Smart Farming
by Mohammed Nagib Hasaneen

Preface

Agriculture is a fundamental sector in the development of any nation. With growing populations, the demand for food has increased, and so has the need for efficient and sustainable agricultural practices. Smart farming is an emerging field that integrates modern technologies and sustainable practices to enhance crop production, reduce environmental impact, and increase farmers' profits. This book explores the latest developments and opportunities in this field.

The introductory chapter provides an overview of smart farming and its significance in the context of sustainable agriculture. It discusses the principles and practices of smart farming and highlights the benefits and challenges of adopting this approach.

Chapter 2 focuses on the adoption of conservation agriculture as a driver of sustainable farming. It examines the opportunities, constraints, and policy issues related to conservation agriculture and how it can contribute to sustainable farming practices.

Chapter 3 explores the role of information technology drivers in smart farming management systems. It discusses the use of sensors, data analytics, and machine learning in precision farming, which can optimize crop yield, reduce resource wastage, and improve sustainability.

Physiological breeding as a smart farming approach is the subject of Chapter 4, which discusses the latest advances in plant breeding techniques that can improve plant resistance to biotic and abiotic stresses, enhance nutrient uptake, and boost productivity.

Chapter 5 focuses on recent advances in nanotechnology, nanomaterials, nanofertilizers, and their applications in smart farming. It discusses the potential benefits of using nanotechnology in agriculture, including improved nutrient uptake, enhanced crop growth, and reduced environmental impact.

Overall, this book provides a comprehensive overview of the latest developments in smart farming and how it can contribute to sustainable agriculture. It is intended for researchers, policymakers, and practitioners in the field of agriculture who are interested in exploring the latest advances in smart farming.

Dr. Subhan Danish
Guest Lecturer,
Faculty of Agricultural Sciences and Technology,
Bahauddin Zakariya University,
Multan, Pakistan

Dr. Hakoomat Ali
Professor,
Faculty of Agricultural Sciences and Technology,
Bahauddin Zakariya University,
Multan, Pakistan

Dr. Rahul Datta
Assistant Professor,
Mendel University,
Brno, Czech Republic

Section 1

Introduction

Chapter 1

Introductory Chapter: Smart Farming

Subhan Danish, Hakoomat Ali and Rahul Datta

1. Introduction

Farming has always been an essential human activity that has sustained civilization for thousands of years. With the rapid growth in population and the consequent demand for food, it has become increasingly important to optimize farming practices to meet the needs of the world's growing population [1]. In recent years, technological advancements have revolutionized the way we approach farming, leading to the emergence of a new approach known as "smart farming" [2]. Smart farming is an innovative approach to agriculture that integrates technology into farming practices, enabling farmers to optimize crop yields, reduce waste, and improve efficiency [3]. This approach uses a range of technologies, including sensors, drones, artificial intelligence, and the internet of things (IoT), to collect data and provide real-time insights into crop health, soil quality, and other key indicators [4]. Smart farming also offers numerous benefits, including increased productivity, reduced labor costs, improved crop quality, and a more

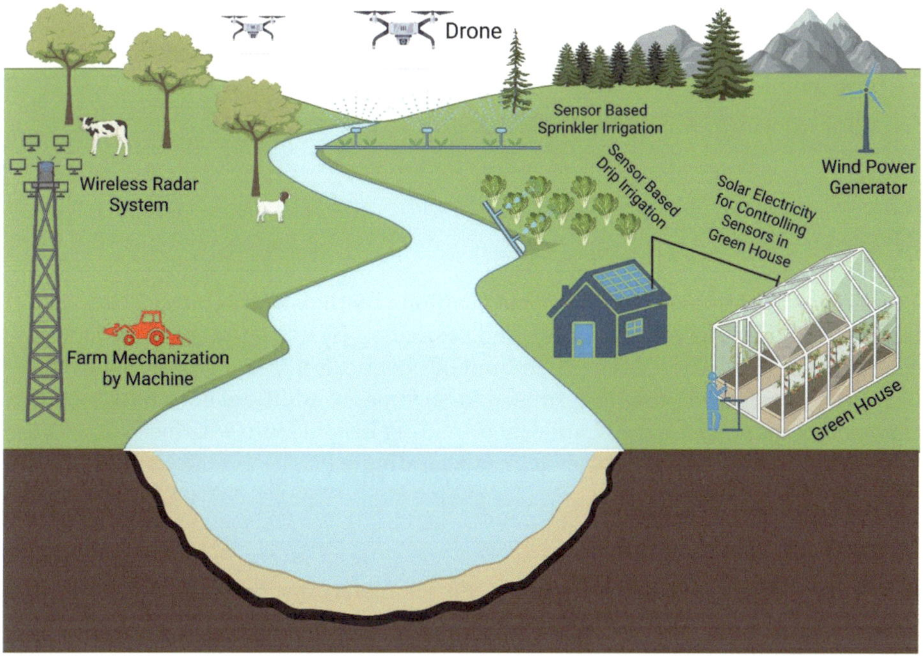

Figure 1.
Smart farming using artificial intelligence on farm for the improvement in agriculture production.

sustainable approach to farming. This approach also offers greater precision, enabling farmers to target specific areas of their farms that require attention and reduce the use of chemicals and fertilizers (**Figure 1**) [5].

2. The evolution of farming: from traditional to smart farming

The evolution from traditional to smart farming can be traced back to the early 1990s when precision agriculture (PA) was first introduced. PA is a farming approach that involves using technology to target specific areas of the farm that require attention, such as soil moisture levels or nutrient deficiencies. This approach uses data analysis tools to optimize inputs and minimize waste, resulting in higher crop yields and reduced costs [1].

Over the years, smart farming has evolved to include a range of advanced technologies. For example, drones equipped with cameras and sensors can provide detailed images and data on crop health and yield. Soil sensors can measure soil moisture, temperature, and nutrient levels, providing insights into the health of the soil and enabling farmers to make informed decisions about fertilization and irrigation [6]. Another technology that has revolutionized smart farming is the IoT. IoT-enabled sensors and devices can be placed throughout the farm to monitor environmental conditions, track crop growth, and optimize irrigation and fertilizer applications. This data is transmitted to a central platform, where it is analyzed and used to generate insights that can help farmers make informed decisions [7].

3. The benefits of smart farming

3.1 Increased productivity

Smart farming allows farmers to collect data on crop health, soil quality, and other key indicators in real time. This data can be analyzed to optimize inputs such as fertilizers, water, and pesticides, resulting in higher crop yields. By targeting specific areas of the farm that require attention, farmers can also reduce waste and ensure that resources are used efficiently [8].

3.2 Sustainability

Smart farming promotes sustainable farming practices by minimizing the use of resources such as water, fertilizers, and pesticides. By using precision agriculture techniques, farmers can reduce the amount of chemicals used on crops, resulting in a more environmentally friendly approach to farming. In addition, smart farming can help farmers adapt to climate change by providing insights into weather patterns and enabling them to adjust farming practices accordingly [9].

3.3 Cost savings

By optimizing inputs and reducing waste, smart farming can lead to significant cost savings for farmers. For example, by using sensors to monitor soil moisture levels, farmers can reduce water usage and save money on irrigation costs. By reducing the use of pesticides and fertilizers, farmers can also save money on these inputs, while also reducing the environmental impact of their farming practices [10].

3.4 Improved crop quality

Smart farming can help farmers improve the quality of their crops by providing insights into crop health and identifying potential issues early on. By using data to optimize inputs and target specific areas of the farm that require attention, farmers can produce higher-quality crops that are more resistant to pests and disease [11].

3.5 Better decision-making

Smart farming provides farmers with real-time data and insights into their farming practices. This data can be used to make informed decisions about inputs, planting schedules, and other factors that can impact crop yields. By using data analysis tools, farmers can also identify trends and patterns that can inform long-term decision-making [12].

4. Challenges to adopting smart farming: costs, training, and infrastructure

While smart farming offers numerous benefits, there are also several challenges that farmers face when adopting this innovative approach to agriculture. Here are some of the main challenges to adopting smart farming:

4.1 Costs

One of the main challenges to adopting smart farming is the cost. Investing in technology such as sensors, drones, and IoT devices can be expensive, particularly for small farmers who may not have the financial resources to invest in this technology. In addition to the initial cost of the technology, there may also be ongoing maintenance and repair costs to consider [13].

4.2 Training

Another challenge to adopting smart farming is the need for specialized training. Farmers need to be trained on how to use the technology, collect and analyze data, and interpret insights. This can be a time-consuming process and may require farmers to take time away from their daily farming activities [12].

4.3 Infrastructure

Smart farming relies on a robust infrastructure to collect and transmit data. This can be a challenge in rural areas where there may be limited access to high-speed internet and other necessary infrastructure. Farmers may need to invest in infrastructure upgrades to support the use of smart farming technology [8].

4.4 Data management

Smart farming generates a large amount of data, and farmers need to have the necessary tools and skills to manage and analyze this data effectively. This can be a challenge for farmers who may not have experience with data analysis or may not have access to the necessary software tools [10].

4.5 Security and privacy

The use of technology in farming raises concerns about data security and privacy. Farmers need to ensure that their data is protected from unauthorized access and that they are complying with relevant data privacy regulations [14].

Despite these challenges, the benefits of smart farming make it an attractive option for farmers looking to increase productivity, reduce waste, and promote sustainability. As technology continues to evolve and become more affordable, it is likely that the adoption of smart farming will continue to grow, enabling farmers to achieve greater efficiency and sustainability.

Keeping in mind the importance of smart farming in our future, this book was planned to provide a comprehensive overview of smart farming, covering topics such as the technologies involved, their applications, and the benefits they offer. We have examined some of the challenges that farmers face when adopting smart farming practices and explore the potential of smart farming to transform the agriculture industry. In this book, we also explore the various aspects of smart farming in greater detail, providing a practical guide for farmers and agricultural professionals seeking to adopt this approach. We hope that the book will inspire more farmers to embrace smart farming and realize its potential to revolutionize the way we produce food.

Author details

Subhan Danish[1]*, Hakoomat Ali[2] and Rahul Datta[3]

1 Faculty of Agricultural Sciences and Technology, Department of Soil Science, Bahauddin Zakariya University, Multan, Punjab, Pakistan

2 Faculty of Agricultural Sciences and Technology, Department of Agronomy, Bahauddin Zakariya University, Multan, Punjab, Pakistan

3 Faculty of Forestry and Wood Technology, Department of Geology and Pedology, Mendel University in Brno, Brno, Czech Republic

*Address all correspondence to: sd96850@gmail.com

References

[1] Wells JCK, Stock JT. Life history transitions at the origins of agriculture: A model for understanding how niche construction impacts human growth, demography and health. Frontiers in Endocrinology. 2020;**11**. Article number: 325

[2] Sharma V, Tripathi AK, Mittal H. Technological revolutions in smart farming: Current trends, challenges and future directions. Computers and Electronics in Agriculture. 2022b;**201**:107217

[3] Said Mohamed E, Belal A, Kotb Abd-Elmabod S, El-Shirbeny MA, Gad A, Zahran MB. Smart farming for improving agricultural management. Egypt. J. Remote Sens. Sp. Sci. 2021;**24**:971-981

[4] Sharma A, Georgi M, Tregubenko M, Tselykh A, Tselykh A. Enabling smart agriculture by implementing artificial intelligence and embedded sensing. Computers and Industrial Engineering. 2022a;**165**:107936

[5] Raj EFI, Appadurai M, Athiappan K. Precision farming in modern agriculture. In: Choudhury A, Biswas A, Singh TP, Ghosh SK, editors. Smart Agriculture Automation Using Advanced Technologies. Singapore: Springer; 2021. pp. 61-87

[6] Boursianis AD, Papadopoulou MS, Diamantoulakis P, Liopa-Tsakalidi A, Barouchas P, Salahas G, et al. Internet of things (IoT) and agricultural unmanned aerial vehicles (UAVs) in smart farming: A comprehensive review. Internet of Things. 2022;**18**:100187

[7] Nolz R, Kammerer G, Cepuder P. Calibrating soil water potential sensors integrated into a wireless monitoring network. Agricultural Water Management. 2013;**116**:12-20

[8] Hrynevych O, Blanco Canto M, Jiménez García M. Tendencies of precision agriculture in Ukraine: Disruptive smart farming tools as cooperation drivers. Agriculture. 2022;**12**:698

[9] Walter A, Finger R, Huber R, Buchmann N. Smart farming is key to developing sustainable agriculture. Proceedings of the National Academy of Sciences. 2017;**114**:6148-6150

[10] Monteiro A, Santos S, Gonçalves P. Precision agriculture for crop and livestock farming—Brief review. Animals. 2021;**11**:2345

[11] Imran MA, Ali A, Ashfaq M, Hassan S, Culas R, Ma C. Impact of climate smart agriculture (CSA) practices on cotton production and livelihood of farmers in Punjab, Pakistan. Sustainability. 2018;**10**:2101. DOI: 10.3390/su10062101

[12] Amadu FO, McNamara PE, Miller DC. Understanding the adoption of climate-smart agriculture: A farm-level typology with empirical evidence from southern Malawi. World Development. 2020;**126**:104692

[13] Thakur D, Kumar Y, Kumar A, Singh PK. Applicability of wireless sensor networks in precision agriculture: A review. Wireless Personal Communications. 2019;**107**:471-512

[14] Gupta M, Abdelsalam M, Khorsandroo S, Mittal S. Security and privacy in smart farming: Challenges and opportunities. IEEE Access. 2020;**8**:34564-34584

Section 2

Sustainable Farming

Chapter 2

Adoption of Conservation Agriculture as a Driver of Sustainable Farming: Opportunities, Constraints, and Policy Issues

Pomi Shahbaz, Shamsheer ul Haq and Ismet Boz

Abstract

Sustainable farming is critical for rural development and global food security, but it is threatened by intensive agriculture and climate change. Conservation agriculture (CA) is a sustainable farming system developed in response to intensive agriculture, environmental degradation, and climate change caused by traditional agriculture systems. This chapter discusses the role of CA in sustainable farming and examines the factors influencing CA adoption globally through a review of previous studies. The review results indicated that CA assists farmers increase farm sustainability by influencing economic, social, and environmental dimensions through minimum mechanical soil disturbance, permanent soil cover, and diversification. CA adoption aims at maintaining soil fertility, improving farm yield, and reducing the use of external inputs necessary for sustainable farming. Therefore, the number of CA-adopting countries has grown significantly over the last decade but its adoption is constrained by a variety of factors such as farmers' demographic characteristics, farm characteristics, institutional factors, capital ownership, cognitive factors, and farm manager entrepreneurial ability. Moreover, abundance of small-scale farming and a lack of awareness about the role of CA in sustainable farming also pose a challenge to the global adoption of CA. Farmers' entrepreneurial abilities and awareness of CA benefits should be improved to increase adoption of CA and sustainable farming.

Keywords: conservation agriculture, farm sustainability, sustainable farming, sustainable land management, sustainable agriculture

1. Introduction

Farming systems play a critical role in ensuring food security worldwide, and healthy soils are necessary for sustainable food production. Farming systems are under huge pressure to meet the increasing demands of agricultural commodities due to an increasing global population and climate change. Farming systems try to meet growing demands through intensified agriculture. Intensified agriculture poses a threat to sustainable farming as it affects the quality of natural resources [1]. Monoculture farming, heavy use of off-farm inputs, and machinery have multifaceted negative impacts on the environment and soil health [2]. The intensive nature of conventional farming raises the risk of soil degradation. Moreover, land and soil degradation have increased dramatically because farming systems have transitioned from high manpower and low input production to low manpower, high use of external inputs, and a highly mechanized system. Machines with more horsepower that move faster than required speed harm the quality and health of the soil. This also increases the loss of soil organic matter, slows water infiltration, and lowers the soil's ability to hold water, all of which are prerequisite for sustainable production [3]. Thus, degradation of land and ecological system services due to intensified agriculture should be avoided, and previous degradation of land must be remedied for sustainable farming [4].

In addition to intensified agriculture, climate change also poses a serious threat to global farming sustainability. Agriculture accounts for nearly one-fourth of total global greenhouse gas emissions and is also directly affected by the effects of climate change [5]. Thus, agriculture is both a cause and an affectee of climate change [6]. Moreover, extreme climatic events are occurring more frequently and are having a severe impact on agriculture by degrading soil and land health [7].

Climate change and land degradation necessitate a more sound and sustainable farming production paradigm that is both environmentally sustainable and economically profitable without compromising yield and productivity [5]. Furthermore, the production system's flexibility and strength must be increased in response to shocks and stress caused by climate change. Similarly, increasing biodiversity above and below ground in the crop production system has numerous important benefits that improve soil health and enable farmers to produce in a way that is supported by society [8]. All of these measures result in sustainable farming, which includes increased production (economic sustainability), a healthier environment, and high resilience to climatic shocks and stress (environmental sustainability) [9]. To conserve and enhance the natural habitats and resources of the environment, "sustainable production intensification" is a new production paradigm [10], which acknowledges the prerequisite for productive and remunerative farming [4, 11]. So, all of these goals can be reached with a no-till method, which is also called Conservation Agriculture (CA). Tillage has a significant impact on soil health because it disrupts the soil's water retention capacity, temperature, and evapotranspiration process [1]. Furthermore, tillage results in a significant loss of soil organic carbon [12].

CA is a sophisticated modern production system that enables farmers to perform sustainable production, which leads to the achievement of sustainable farming goals [13]. Somasundaram et al. [14] define "CA" as "a set of management practices for sustainable agricultural production that avoids excessive soil disturbance in order to protect it from soil degradation processes such as erosion, compaction, structural/aggregate breakdown, loss of soil organic matter, and nutrient leaching. FAO [15] defines CA as a farming system that promotes minimum soil disturbance, the maintenance of a permanent soil cover, and the diversification of plant species. Thus, CA has three principles, which are as follows: 1) low mechanical soil disturbance; 2) permanent soil cover; and 3) diversification (**Figure 1**).

No-tillage, minimal disturbance, and direct seeding without tilling the soil are all examples of minimal mechanical soil disturbance. It explains how to cause the least amount of soil disturbance through cultural farm practices or mechanical operations. Direct seeding into soil is encouraged for sustainable farming [16]. FAO [15] also suggested that the disturbed area should be 15 cm wide or less than 25% of the total cropped area. Minimal or no-tillage is an effective erosion control measure that increases fertilizer and water use efficiency and crop yield [17].

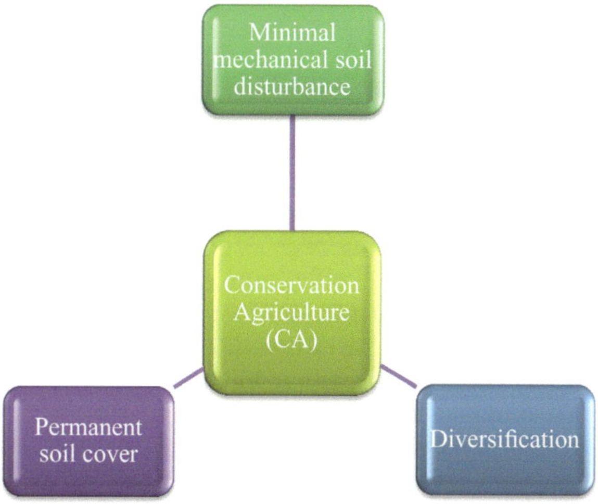

Figure 1.
Principles of conservation agriculture.

Permanent organic soil cover refers to the ground surface's permanent biomass soil mulch cover. It is particularly encouraged when there is a long gap between harvesting and planting the next crop [15]. Crop biomass, cover crops, and rootstocks can all be preserved. Microbes decompose the cover crop naturally in the soil [18]. When the field is empty, it protects the soil and mobilizes and accelerates the nutrient recycling process. Soil cover also preserves the soil structure and reduces hardpans and compacted layers. Moreover, it also reduces weed growth and pest attacks.

Diversification refers to the preservation of soil nutrients through crop rotation, which entails the proper sequence and association of annual and perennial crops, as well as a balanced mix of legume and nonlegume crops [5]. Crop rotation feeds the soil because many nutrients leach down to the deeper layers of soil and are no longer available for better crop growth. These nutrients are naturally recycled through proper and balanced crop rotation [15].

As a result, CA is critical for sustainable farming, and this chapter will discuss CA's brief history and current global situation. Furthermore, the chapter will discuss the role of CA in sustainable farming and list the CA practices that are being implemented on farms around the world. The chapter also aims to provide information on the challenges that farmers face when implementing different farm practices on their farms and concludes with policy recommendations for improving the CA situation, especially in developing countries.

2. History and global status of CA adoption

2.1 Historical background of CA

Tillage is the use of farm machinery to manipulate soil. Tillage has a long history, dating back a million years, when men transitioned from hunting to sedentary and conventional farming, particularly in the Nile, Euphrates, Tigris,

Indus Valley, and Yangtse valleys [19]. Tillage was traditionally used to soften the soil layers for seedbed preparation, control and manage weeds, and improve the oxidation mineralization process [13]. In the years following the industrial revolution in the 1990s, the invention of the engine made machinery available for performing farm activities, such as plowing, planking, seed drilling, and so on. In the Midwestern United States, dust bowls destroyed large areas, and tillage-based farming was called into question for the first time in history in the 1930s [20]. As a result, for the first time, CA practices such as reducing tillage and covering the soil for soil protection were adopted on farms. The seedling machine was invented in the 1990s, allowing seeds to be planted without disturbing the soil. CA was first theoretically proposed in 1943 by Edward H. Faulkner in the manuscript "Plowman's Folly" [21]. The CA idea has become more and more popular over time and is used a lot in sustainable farming.

No-tillage was first used in farming in Brazil in the early 1970s, and agriculture has been transformed by incorporating no-tillage practices into the farming system known as CA today. Furthermore, in the 1970s, no-tillage was practiced in West Africa [22, 23]. Before CA reached a significant adoption level in South America and the rest of the world, significant improvements in farm equipment and agronomic practices regarding CA were made and developed to enhance crop growth and machine efficiency. Also, as fuel prices went up in the 1970s, farmers switched to a system that farming resources. Commercial farmers used the CA to avoid soil erosion caused by drought, along with the fuel-saving system [24].

Since the early 1990s, the CA has become well known and has spread rapidly, and agricultural systems in Brazil, Paraguay, and Argentina have been transformed into CA. The development of the CA system drew the attention of the rest of the world, and international organizations such as FAO, CGIAR, IFAD, EU, ACT, CIRAD centers, and many others began to take an interest in the CA system's promotion. Following that, a study tour to Brazil, research projects, and workshops were organized all over the world to raise awareness and increase CA adoption. After that, CA adaptation has been observed in African countries such as Tanzania, Zambia, and Kenya, as well as in Asia, particularly in China, Pakistan, India, and Kazakhstan. The CA was also significantly adapted in developed countries such as Australia, Spain, Canada, and Finland at the end of the millennium [25]. The CA adaptation is not restricted to specific geographical and ecological environments. Farmers practice it from the Arctic Circle (e.g., Finland) to the tropics (e.g., Uganda, Kenya). CA has also been adopted at 3000 m altitude and under severe environmental conditions with 250 mm of rain a year (e.g., Morocco, Western Australia), as well as in countries such as Brazil and Chile where heavy rainfall occurs during the whole year. No-tillage is also used in sandy and clay soil types. It is used in soils ranging from 90% sand (as in Australia) to 80% clay (as in Brazil's Oxisols and Alisols). Similarly, the CA system can grow any crop [20, 26].

2.2 Global situation of CA adoption

CA is currently practiced in over 79 countries worldwide [5], and the number of farmers adopting CA practices on farms is increasing in both developed and developing countries due to its beneficial effects on farm resources and crop production. CA is practiced on every continent, but Europe has the most countries that have adopted

CA (**Figure 2**). CA has also begun to gain traction in Asia and Africa, with the number of countries adopting CA on these two continents increasing significantly over the last decade.

Despite the fact that the number of countries adopting CA has increased significantly, the area under CA remains minimal in comparison to the total world cropped area. In 2015/16, the total area under CA in the world was 180.44 million hectares. Europe has a large majority of CA-adopting countries, but its share of the total global CA area is negligible. Similarly, Asia and Africa contain significant world agricultural land as well as habitats for the world's large population, which is more vulnerable to climate change and food insecurity, but their share of total CA world area is also minute.

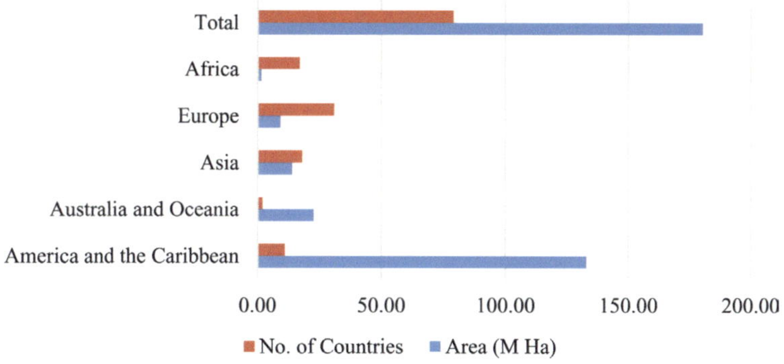

Figure 2.
Area under CA and number of adopting countries on each continent.

Continent/country	CA area share (%)
America and the Caribbean	
United States	32.47
Brazil	24.05
Australia and Oceania	
Australia	98.39
New Zealand	1.61
Asia	
China	64.61
Kazakhstan	17.95
Europe	
Russia	54.01
Spain	9.72
Africa	
South Africa	29.09
Zambia	20.94

Table 1.
Countries with the highest share in area under CA in their continents.

The United States and the Caribbean countries that practice CA have the largest share of the total CA area in the world, followed by Australia and Oceania. **Table 1** depicts the major CA adopting countries by area on each continent. The United States accounts for nearly one-third, with Brazil accounting for nearly one-fourth of the total area under CA in America and the Caribbean countries. Australia is the largest CA adopter on the Australian continent. CA was widely practiced in Asia, with China accounting for nearly two-thirds of the total area covered, followed by Kazakhstan. Russia and Spain were major contributors to the CA area in Europe. South Africa alone accounts for nearly one-third of the total area under CA on the African continent.

3. Conservation agriculture and sustainable farming

Sustainable development is defined as "the ability of the current generation to meet their needs without jeopardizing future generations' ability to meet their own needs" [27]. In order to apply the concept of sustainable development to farming, farming must be socially acceptable, economically viable, and environmentally friendly. Farming will therefore be sustainable if it is socially, economically, and environmentally sustainable [28]. Hobbs et al. [13] described CA as a modern way of farming that helps farmers achieve their goals of sustainable development through sustainable farming.

The impact of CA on crop yield can be used to explain CA's role in achieving economic sustainability [29, 30]. A high crop yield is one of the farmer's primary goals in order to enjoy a good economic return. Crop yield is affected by numerous management (timely and proper application of off-farm inputs) and ecological factors (uneven rains, harsh weather, water deficiency, etc.). CA is critical in reducing the negative impact of these factors on crop yield. No-tillage, for example, improves soil fertility and structure while also softening the soil, which improves seed germination and crop growth [31]. A well-grown crop with good germination results in a good crop yield, which increases the crop surplus. Farmers make a good profit from the high crop surplus, which lets farmers keep their high standard of living. CA also boosts crop yield and economic returns, which makes farming socially and economically more sustainable. Thus, the CA improves farmers' long-term welfare by increasing crop yield, high economic returns, and food security [32–35].

Zheng et al. [36] discovered that adoption of CA practices has a positive impact on crop yield in China. The study showed that adoption of CA on farms significantly increases crop yields. However, the impact of CA on crop yields is dependent on geographical location, climatic conditions, and the type of adopted CA practice. The study also found that conventional tillage with straw retention produced better crop yields than no-tillage with straw retention. Moreover, the study also reported that CA practices produce better results in geographical locations with annual precipitation of less than 600 mm and a mean temperature of greater than 5°C. The potential for high crop yields with CA is greater in rain-fed areas than in conventional tillage systems [37]. The CA is more effective in terms of yield and farm productivity effects when all three CA principles are implemented in combination than when they are implemented alone on farms. Even sometimes, CA principles implemented separately can have a negative impact on farm productivity. Pittelkow et al. [29] stated that no-tillage has a negative impact on crop yield, but when combined with the other two principles (cover crops and crop rotation); it generates an equal or greater crop yield than conventional agriculture. Therefore, no-tillage, cover crop, and diversification (crop rotation) produce high crop yields ensuring the CA system's economic sustainability [38–40].

The widespread adoption of three CA principles around the world has ushered in a new era of environmental control and mitigation for damages associated with conventional agriculture. No-tillage leaves the soil untouched, improving its physical properties, which is a major component of the environment. The organic carbon stock is three times more concentrated in the soil than in the atmosphere [31, 41]. Increased soil carbon level is highly associated with increased soil carbon level through improved mineralization processes, which reduces the negative effects of climate change on crop yield [42, 43]. CA provides climate-smart sustainable farming systems that enable farmers to cope with the adverse impacts of climate change [44]. Similarly, growing cover crops on fallow lands lowers the risk of soil erosion [45]. By covering the field where no primary crop is grown, it can control weed germination and enhance N-input leaching [46]. It also increases the soil's water-holding capacity, promotes microbial activity, maintains soil structure and porosity, and balances the nutrient cycle [47–50]. Similarly, crop rotation reduces climate vulnerability and improves soil health by reducing herbivores, increasing yields, and generating high economic benefits. It also provides a more stable planting system in extreme weather conditions [51–53]. Likewise, CA improves soil fertility [54], reduces soil erosion [55], improves water filtration and retention, and reduces greenhouse gas emissions [56], all of which contribute to the farming system's high environmental sustainability.

Therefore, the CA is an ideal solution for resolving the environmental problem in agriculture. CA's associated environmental benefit improves agriculture's environmental sustainability.

Moreover, CA also contributes to social sustainability by increasing gender equality, labor participation, and farmer welfare, and it is expected that promoting the CA farming system will increase the participation of women in farming [57].

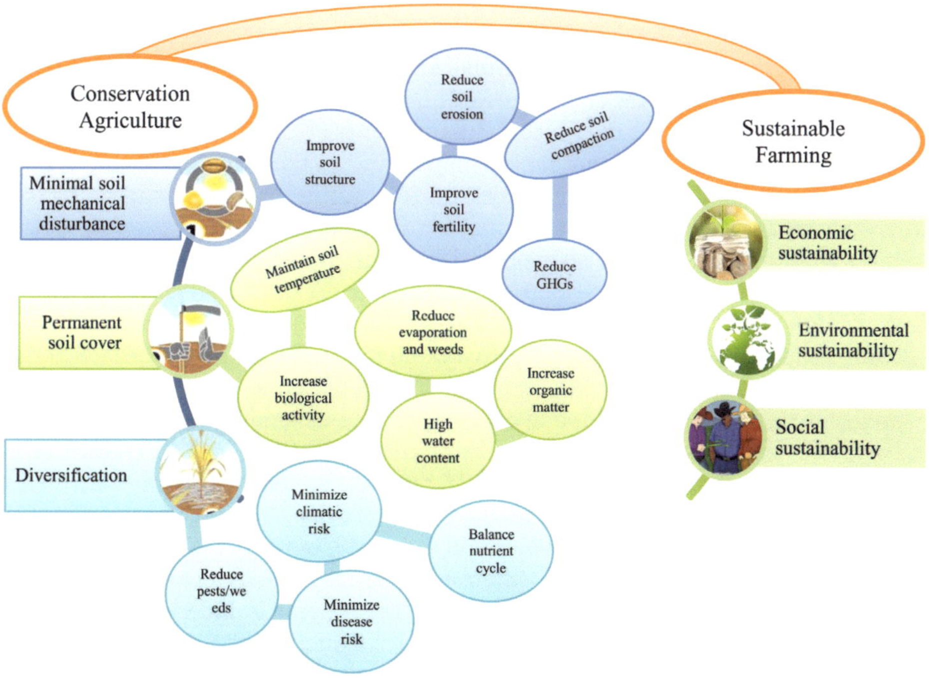

Figure 3.
Conservation Agriculture (CA) and Sustainable Farming (SF).

For example, women's participation in Zimbabwe increased grain yield and improved food diversity and security. Furthermore, the use of CA has altered intra-household decision-making between males and females. Women's participation in decision-making, crop management, and improved agency were observed in Zimbabwe [58, 59]. Furthermore, crop residue retention can be used to describe the labor requirements of CA [60]. Women practicing CA are good time managers because they start clearing land on time to prepare it for early planting [61].

Based on the preceding discussion, it is clear that CA adoption helps farmers cope with climate change while also increasing farm-level sustainability (social, economic, and environmental). Thus, adoption of CA works as a driver of sustainable farming, as depicted in **Figure 3**. CA principles provide multifaceted benefits that lead to sustainable farming. Improving soil fertility, for example, improves farmers' economic conditions, which in turn affects farms' economic and social sustainability. Similarly, reducing the use of machinery for tillage reduces greenhouse gas emissions while also reducing costs, which helps to improve environmental, economic, and social sustainability. As an outcome, CA's minimal soil mechanical disturbance principle influences all three dimensions of sustainability. Similarly, the other two CA principles help farmers improve their farms' economic, social, and environmental sustainability. As a result, CA adoption can play a critical role in global farming sustainability.

4. Globally adopted CA practices and constraints

4.1 CA practices and factors influencing CA adoption on farms

CA practices aim to improve farm resource utilization by integrating natural resource management such as soil, water, and biological resources with the fewest external farm inputs [62]. Therefore, CA is being adopted all over the world in response to the growing concerns of national and international institutions related to farm sustainability, and CA is one of the most important and rapidly expanding adoption systems in all regions of the world. Different types of CA practices are preferred on different farms and in different regions depending on the climate, the land type, the farmer's skills, and the farm's resources. Moreover, the CA practices adoption on farms is also majorly dependent on the purpose of adoption CA. Moreover, the CA practices adoption on farms is also majorly dependent on the purpose of adoption. As a result, CA practices used in one country or farm may be different from those used in another country due to the difference in intended objectives of CA adoption.

Table 2 shows the CA practices adopted around the world, as well as the factors that influence the adoption of these CA practices on farms. Different types of CA practices are preferred on different farms and in different regions depending on the climate, the land type, the farmer's skills, and the farm's resources. According to the literature, farmers in different countries adopted different CA practices on their farms, but zero tillage/no-tillage was one of the most widely adopted CA practices on farms. The minimum or zero-tillage CA practice is widely used around the world, but crop residuals retention in the field is more complicated [74]. Moreover, cover crops and crop rotation are also commonly practiced CA strategies in the world. Crop rotation is underutilized in terms of pest control, disease cycle disruption, income risk reduction, and soil fertility [74].

Farmers' demographic characteristics, farm characteristics, institutional and social inclusion, capital ownership, and cognitive factors (farmer attitude, CA perception, and farming behavior) were identified as major influencing factors in

Country/Region	CA practices	Influencing factors	Objective/s
Nigeria [63]	Zero tillage, minimum tillage, contour stripping, not burning field, tree planting, cover crops, dead tree trunks, mulching	Age, education, innovativeness, attitude toward conservation, risk bearing, credit, farm income, input-output prices, off-farm occupation	Soil erosion controlling
Rwanda [64]	Organic inputs	Monetary and physical factors, human capital, investment risk, wealth, and liquidity sources	Land conservation investment and organic inputs use
United States [65]	Conservation tillage, contour farming, strip-cropping, grass waterways	Farm size, age, college education, program participation, land tenure, annual precipitation.	Adoption of CA practices and land tenure
Zimbabwe [66]	Zero tillage, crop rotation, contour ridging technologies	Farm and farmer characteristics, institutional factors	Analyzing the adoption of CA by small farmers
Spain [67]	Not burning olive-de suckering debris, using shredded olive-pruning debris as soil cover, cover crops under mower control	Socio-economic characteristics of a farmer, social capital indicators, farm characteristics, farm management	Assessing the soil conservation practice as CA in olive groves
Zimbabwe [68]	Winter weeding, digging planting basins, crop residues, manuring, basal fertilizer, topdressing, Timely weeding, crop rotation	Age, Education, own land, draught power, Extension services, labor, conservation agricultural experiences.	Assessing the adoption of CA among different clusters of farmers.
Australia [69]	No-tillage, crop stubble, legumes rotation, controlled traffic farming,	Various socio-economic factors	Assessing the adoption of CA as climate change mitigation activity
Bangladesh [70]	Conservation agriculture principles	Farm size, family size, farming experience, age and education of head, extension services, farm and off-farm income	Impact of CA adoption farms' economic viability
Kenya [71]	Mulching, direct planting, shallow weeding, spraying herbicides	Attitudes, perceived norms and perceived behavioral control, farmer's perception Of the social norms towards CA, farmer's perceived behavioral control	Assessing the farmer's decision of selecting the CA over conventional agriculture.
South Africa [72]	No-till conservation agriculture	Age, gender, education, experience, training, extension, credit access, land size, income	Assessing the adoption of no-till conservation by small-scale farmers
Malawi [35]	Zero tillage, mulching of crop residual, and intercropping with legumes	Landholding, education, neighbor's adaption of CA, gender, no. of male and females in family	Decision-making analysis regarding CA adaptation
United States [73]	Conservation tillage, cover crops, diverse crop rotation	Location and spatial variable, age, college education, area under operation, family network index, organization network index, perception of environmental benefits of the practices.	Assessing the adaption of CA practices

Table 2.
CA practices around the world and the factor influencing its adoption.

the adoption of CA practices worldwide. Age, education, farming experience, and family size were among the socio-demographic factors influencing the adoption of CA practices on farms. Farm size, farm income, and land tenure status were the farm characteristics that influenced CA adoption in various countries. Credit utilization and the availability of extension services were institutional determinants of CA adoption. Furthermore, famers' attitudes and perceptions were influential factors in the global adoption of CA practices. Giller et al. [75] also stated that socioeconomic factors play an important role in CA practice adaptation.

4.2 Challenges in adoption of CA practices on farms

Although CA is a driving force in achieving sustainable farming, there are numerous challenges and constraints that affect CA adoption on farms around the world, in addition to the factors discussed above. The first and most prominent challenge that limits the adoption of CA cited in previous literature is small-scale farming. Small farmers lack social inclusion, institutional support, capital, and other resources, and they are less likely to implement CA on their farms. Moreover, farmers with owned animal traction are also less likely to adopt CA practices such as minimum tillage/zero tillage on their farms due to readily available tillage sources. Framers with animal traction typically replace mechanized soil management with animal-driven plowing [76].

Crop residual retention is also used as a CA practice in the fields by the farmers. Crop cultivation and livestock rearing are complementary to each other for farmers, especially small farmers in developing countries, but farmers with crop and livestock interaction tend to have low crop residual retention on their farms. Furthermore, livestock is a more important source of traction and income security in an emergency [76], and crop residuals are a vital source of animal feed [77]. Furthermore, managing crop residuals is more expensive than simply burning crop residuals. The cost of managing crop residuals is more than one-third higher than the cost of burning the residuals [78, 79].

Another major impediment to the adoption of CA practices such as crop rotation is a lack of timely seed availability, as well as dysfunctional markets for final farm outputs [80]. The other challenge faced by farmers in the adoption of CA practices on farms is the unavailability of advanced equipment required for CA adoption. The equipment is either unavailable or its financial costs are high, especially in developing countries. Other than the above challenges, lack of awareness about different CA practices and their associated benefits among farmers in developing countries limits the adoption of CA on farms.

5. Conclusion

Agriculture is one of the largest consumers of natural resources, and adopting sustainable farming systems is necessary not only to preserve natural resources but also to meet the food needs of the world's ever-increasing population under climatic change scenarios. A sustainable farming system is one that is socially acceptable, economically viable, and environmentally friendly. CA is a sustainable system that helps farmers improve farm sustainability by influencing economic, social, and environmental dimensions through its three basic principles. These three principles of CA are as follows 1) minimal mechanical soil disturbance; 2) permanent soil cover; and 3) diversification. The CA system is intended to improve and maintain soil fertility while reducing the use of external farm inputs. Thus, the CA system increases crop yields while decreasing input costs, affecting farmers' economic and social sustainability. Similarly, CA practices aid in mitigating the effects of climate change on farms and also reduce the use of machinery and chemicals in the fields. All of this contributes to the farmers' environmental sustainability. As a result, implementing CA practices on farms is critical not only for the sustainable management of agricultural land but also for the overall sustainability of the farming system.

The European continent has the most CA-adopting countries, and the United States is the world's largest CA adopter in terms of area. Despite the fact that CA is being adopted in many countries around the world, the area under CA remains very small in comparison to the world's total cultivable area. Farmers' demographics, farm characteristics, institutional and social inclusion, capital ownership, and cognitive factors (farmer attitude, CA perception, and farming behavior) were identified as major influencing factors in the global adoption of CA practices. Furthermore, small-scale farming as well as a lack of awareness about the benefits of CA is regarded as major barriers to CA adoption worldwide.

CA can also play an important role in achieving the Sustainable Development Goals in developing countries, as agriculture is a major driver of their economies. The continuous degradation of natural resources, particularly land, is the primary cause of unsustainable farming systems all over the world. As a result, developing countries should increase the adoption of CA practices in order to transition from unsustainable to sustainable farming systems. The following suggestions are recommended for increasing CA adoption in developing countries:

In order for CA to be adopted in developing countries, farms must be treated as enterprises like any other business. So, farmer entrepreneurship and a culture of entrepreneurship in farming should be encouraged to enhance the adoption of CA and sustainable farming by involving all agricultural socio-economic networks (farmers and their associations, farmer cooperatives, research and advisory organizations, market and chain parties, and government and social agencies).

Increasing farmer awareness about the benefits of the CA farming system is critical to the adoption of CA practices on farms in developing countries. The agricultural extension system remains an important source of information for farmers. As a result, agriculture extension systems can play an important role in the adoption of CA and sustainable farming in developing countries by creating awareness among the farming community. Through their agricultural extension network, policymakers need to develop a comprehensive plan for educating farmers, particularly small-scale farmers, about various CA practices and the benefits associated with them. Before providing information to farmers, the first step should be to train extension workers who are directly involved with farmers to improve their own knowledge of CA. Furthermore, in order to achieve sustainable farming goals, the agricultural extension network must be expanded in terms of capacity and outreach to the larger farming community, particularly small farmers and those living far from city centers. Modern information and communication channels should also be used to raise farmers' awareness of the benefits of CA practices and sustainable agriculture along with traditional information systems.

Furthermore, small farmers must be provided with CA equipment for the adoption of CA practices such as zero tillage machines, as the majority of farmers in the developing world have subsistence land sizes.

Finally, engaging the younger generation in agriculture is critical to sustainable farming in both the developing and developed worlds, as youth interest in agriculture as a career has declined dramatically. Therefore, policymakers should look for ways to increase youth participation in agricultural activities in order to ensure sustainable farming.

Conflict of interest

The authors declare no conflict of interest.

Author details

Pomi Shahbaz[1*], Shamsheer ul Haq[2] and Ismet Boz[1]

1 Department of Agricultural Economics, Ondokuz Mayis University, Samsun, Turkey

2 Department of Economics, Division of Management and Administrative Science, University of Education, Lahore, Pakistan

*Address all correspondence to: pomi1781@gmail.com

References

[1] Busari MA, Kukal SS, Kaur A, Bhatt R, Dulazi AA. Conservation tillage impacts on soil, crop and the environment. International Soil and Water Conservation Research. 2015;**3**(2):119-129

[2] Altieri M. Modern Agriculture: Ecological impacts and the possibilities for truly sustainable farming. Agroecology in Action (internet). Available from: http://nature.berkeley.edu/~agroeco3/index.html [2022.04.10]

[3] Dumansky J, Reicosky DC, Peiretti RA. Pioneers in soil conservation and conservation agriculture. International Soil and Water Conservation Research. 2014;**2**(1):1-4

[4] Kassam AH, Basch G, Friedrich T, Shaxson F, Goddard T, Amado T, et al. Sustainable soil management is more than what and how crops are grown. In: Lal R, Stewart BA, editors. Principles of Soil Management in Agro-Ecosystems. Advances in Soil Science. Raton, FA: CRC Press; 2013. pp. 230-270

[5] Kassam A, Friedrich T, Derpsch R. Global spread of conservation agriculture. International Journal of Environmental Studies. 2019;**76**(1):29-51

[6] Haq SU, Boz I, Shahbaz P. Adoption of climate-smart agriculture practices and differentiated nutritional outcome among rural households: A case of Punjab province, Pakistan. Food Security. 2021;**13**(4):913-931

[7] Hay J. Extreme weather and climate events, and farming risks. In: Managing Weather and Climate Risks in Agriculture. Berlin, Heidelberg: Springer; 2007. pp. 1-19

[8] Brussaard L, De Ruiter PC, Brown GG. Soil biodiversity for agricultural sustainability. Agriculture, Ecosystems and Environment. 2007;**121**(3):233-244

[9] Montgomery CA. Ranking the benefits of biodiversity: An exploration of relative values. Journal of Environmental Management. 2002;**65**(3):313-326

[10] FAO. Save and Grow, a Policymaker's Guide to Sustainable Intensification of Smallholder Crop Production. Rome: FAO; 2011

[11] Goddard T, Zoebisch MA, Gan YT, Ellis W, Watson A, Sombatpanit S. No-Till Farming Systems. Bangkok, Thailand: WASWAC;

[12] Lal R. Soil carbon sequestration to mitigate climate change. Geoderma. 2004;**123**(1-2):1-22

[13] Hobbs PR, Sayre K, Gupta R. The role of conservation agriculture in sustainable agriculture. Philosophical Transactions of the Royal Society, B: Biological Sciences. 2008;**363**(1491):543-555

[14] Somasundaram J, Sinha NK, Dalal RC, Lal R, Mohanty M, Naorem AK, et al. No-till farming and conservation agriculture in South Asia–issues, challenges, prospects and benefits. Critical Reviews in Plant Sciences. 2020;**39**(3):236-279

[15] FAO. What is conservation agriculture (internet). 2022. Available from https://www.fao.org/conservation-agriculture/overview/conservation-agriculture-principles/en/ [2022.05.20]

[16] Nichols V, Verhulst N, Cox R, Govaerts B. Weed dynamics and

conservation agriculture principles: A review. Field Crops Research. 2015;**183**:56-68

[17] Triplett GB Jr, Dick WA. No-tillage crop production: A revolution in agriculture! Agronomy Journal. 2008;**100**:153

[18] LFDFN. Permanent Soil Cover: Cover Crops, Lower Fox Demonstration Farms Network (ntrenet). Available from: https://fyi.extension.wisc.edu/foxdemofarms/conservation-agriculture/permanent-soil-cover-cover-crops/#:~:text=Permanent%20soil%20cover%20is%20defined,planted%20late%20in%20the%20season [2022.05.22]

[19] Hillel D. Out of the Earth: Civilization and the Life of the Oil. California: University of California Press; 1992

[20] Friedrich T, Derpsch R, Kassam A. Overview of the Global Spread of Conservation Agriculture. 2012

[21] Faulkner EH. Plowman's Folly. London: Michael Joseph Ltd.; 1943

[22] Greenland DJ. Bringing the Green Revolution to the Shifting Cultivator: Better seed, fertilizers, zero or minimum tillage, and mixed cropping are necessary. Science. 1975;**190**(4217):841-844

[23] Lal R. No-tillage effects on soil properties under different crops in Western Nigeria. Soil Science Society of America Journal. 1976;**40**(5):762-768

[24] Haggblade S, Tembo G. Conservation Farming in Zambia EPTD. Washington: IFPRI;

[25] Farooq M, Siddique KH. Conservation agriculture: Concepts, brief history, and impacts on agricultural systems. In: Conservation Agriculture. Cham: Springer; 2015. pp. 3-17

[26] Derpsch R, Friedrich T. Development and Current Status of No-till Adoption in the World. In: Proceedings on CD, 18th Triennial Conference of the International Soil Tillage Research Organization (ISTRO). Izmir. 2009

[27] Ul Haq S, Boz I. Measuring environmental, economic, and social sustainability index of tea farms in Rize Province, Turkey. Environment, Development and Sustainability. 2020;**22**(3):2545-2567

[28] Ul Haq S, Boz I. Developing a set of indicators to measure sustainability of tea cultivating farms in Rize Province, Turkey. Ecological Indicators. 2020;**95**:219-232

[29] Choi SH, Kim YH, Hebisch M, Sliwinski C, Lee S, D'Avanzo C, et al. A three-dimensional human neural cell culture model of Alzheimer's disease. Nature. 2014;**515**(7526):274-278

[30] Sun W, Canadell JG, Yu L, Yu L, Zhang W, Smith P, et al. Climate drives global soil carbon sequestration and crop yield changes under conservation agriculture. Global Change Biology. 2020;**26**(6):3325-3335

[31] Blanco-Canqui H, Ruis SJ. No-tillage and soil physical environment. Geoderma. 2018;**326**:164-200

[32] Baudron F, Corbeels M, Monicat F, Giller KE. Cotton expansion and biodiversity loss in African savannahs, opportunities and challenges for conservation agriculture: A review paper based on two case studies. Biodiversity and Conservation. 2009;**18**(10):2625-2644

[33] Makate C, Makate M, Mango N. Sustainable agriculture practices and

livelihoods in pro-poor smallholder farming systems in southern Africa. African Journal of Science, Technology, Innovation and Development. 2017;**9**(3):269-279

[34] Nkala P, Mango N, Zikhali P. Conservation agriculture and livelihoods of smallholder farmers in Central Mozambique. Journal of Sustainable Agriculture. 2011;**35**(7):757-779

[35] Ward PS, Bell AR, Droppelmann K, Benton TG. Early adoption of conservation agriculture practices: Understanding partial compliance in programs with multiple adoption decisions. Land Use Policy. 2018;**70**:27-37

[36] Zheng C, Jiang Y, Chen C, Sun Y, Feng J, Deng A, et al. The impacts of conservation agriculture on crop yield in China depend on specific practices, crops and cropping regions. The Crop Journal. 2014;**2**(5):289-296

[37] Farooq M, Flower KC, Jabran K, Wahid A, Siddique KH. Crop yield and weed management in rainfed conservation agriculture. Soil and Tillage Research. 2011;**117**:172-183

[38] Lu YC, Watkins KB, Teasdale JR, Abdul-Baki AA. Cover crops in sustainable food production. Food Reviews International. 2000;**16**(2): 121-157

[39] Volsi B, Higashi GE, Bordin I, Telles TS. Production and profitability of diversified agricultural systems. Anais da Academia Brasileira de Ciências. 2021;**93**(2):1-15

[40] Singh J, Wang T, Kumar S, Xu Z, Sexton P, Davis J, et al. Crop yield and economics of cropping systems involving different rotations, tillage, and cover crops. Journal of Soil and Water Conservation. 2021;**76**(4):340-348

[41] Sanderman J, Hengl T, Fiske GJ. Soil carbon debt of 12,000 years of human land use. Proceedings of the National Academy of Sciences of the United States of America. 2017;**114**(36):9575-9580

[42] Soussana JF, Lutfalla S, Ehrhardt F, Rosenstock T, Lamanna C, Havlík P, et al. Matching policy and science: Rationale for the '4 per 1000-soils for food security and climate' initiative. Soil and Tillage Research. 2019;**188**:3-15

[43] Lobell DB, Schlenker W, Costa-Roberts J. Climate trends and global crop production since 1980. Science. 2011;**333**(6042):616-620

[44] Beuchelt TD, Badstue L. Gender, nutrition-and climate-smart food production: Opportunities and trade-offs. Food Security. 2013;**5**(5):709-721

[45] Battany MC, Grismer ME. Rainfall runoff and erosion in Napa Valley vineyards: Effects of slope, cover and surface roughness. Hydrological Processes. 2000;**14**(7):1289-1304

[46] Halde C, Gulden RH, Entz MH. Selecting cover crop mulches for organic rotational no-till systems in Manitoba, Canada. Agronomy Journal. 2014;**106**(4):1193-1204

[47] Lotter DW, Seidel R, Liebhardt W. The performance of organic and conventional cropping systems in an extreme climate year. American Journal of Alternative Agriculture. 2003;**18**(3):146-154

[48] Drinkwater LE, Snapp S. Nutrients in agroecosystems: Rethinking the management paradigm. Advances in Agronomy. 2007;**92**:163-186

[49] Haruna SI, Nkongolo NV. Cover crop management effects on soil physical and biological properties. Procedia Environmental Sciences. 2015;**29**:13-14

[50] Abdalla M, Hastings A, Cheng K, Yue Q, Chadwick D, Espenberg M, et al. A critical review of the impacts of cover crops on nitrogen leaching, net greenhouse gas balance and crop productivity. Global Change Biology. 2019;**25**(8):2530-2543

[51] Bowles TM, Mooshammer M, Socolar Y, Calderón F, Cavigelli MA, Culman SW, Deen W, Drury CF, y Garcia AG, Gaudin AC, Harkcom WS. Long-term evidence shows that crop-rotation diversification increases agricultural resilience to adverse growing conditions in North America. One Earth. 2020; 2(3):284-293.

[52] Li J, Huang L, Zhang J, Coulter JA, Li L, Gan Y. Diversifying crop rotation improves system robustness. Agronomy for Sustainable Development. 2019;**39**(4):1-3

[53] Yu T, Mahe L, Li Y, Wei X, Deng X, Zhang D. Benefits of crop rotation on climate resilience and its prospects in China. Agronomy. 2022;**12**(2):436

[54] Palm C, Blanco-Canqui H, Declerck F, Gatere L, Grace P. Agriculture, ecosystems and environment. Agriculture, Ecosystems and Environment. 2014;**187**:87-105

[55] Johansen C, Haque M, Bell R, Thierfelder C, Esdaile R. Conservation agriculture for small holder rainfed farming: Opportunities and constraints of new mechanized seeding systems. Field Crops Research. 2012;**132**:18-32

[56] Govaerts B, Verhulst N, Castellanos-Navarrete A, Sayre KD, Dixon J, Dendooven L. Conservation agriculture and soil carbon sequestration: Between myth and farmer reality. Critical Reviews in Plant Sciences. 2009;**28**(3):97-122

[57] Farnworth CR, Baudron F, Andersson JA, Misiko M, Badstue L, Stirling CM. Gender and conservation agriculture in East and Southern Africa: Towards a research agenda. International Journal of Agricultural Sustainability. 2016;**14**(2):142-165

[58] Hove M, Gweme T. Women's food security and conservation farming in Zaka District-Zimbabwe. Journal of Arid Environments. 2018;**149**:18-29

[59] Kunzekweguta M, Rich KM, Lyne MC. Factors affecting adoption and intensity of conservation agriculture techniques applied by smallholders in Masvingo district, Zimbabwe. Agrekon. 2017;**56**(4):330-346

[60] Nyanga PH, Johnsen FH, Kalinda TH. Gendered impacts of conservation agriculture and paradox of herbicide use among smallholder farmers. International Journal of Technology and Development Studies. 2012;**3**(1):1-24

[61] Wekesah FM, Mutua EN, Izugbara CO. Gender and conservation agriculture in sub-Saharan Africa: A systematic review. International Journal of Agricultural Sustainability. 2019;**17**(1):78-91

[62] Garcıa-Torres L, Benites J, Martınez-Vilela A, Holgado-Cabrera A. Conservation Agriculture: Environment, Farmers Experiences, Innovations, Socio-economy, Policy. Boston: Kluwer Academic Publishers;

[63] Okoye CU. Comparative analysis of factors in the adoption of traditional and recommended soil erosion control practices in Nigeria. Soil and Tillage Research. 1998;**45**(3-4):251-263

[64] Clay D, Reardon T, Kangasniemi J. Sustainable intensification in the

highland tropics: Rwandan farmers' investments in land conservation and soil fertility. Economic Development and Cultural Change. 1998;**45**(2):351-378

[65] Soule MJ, Tegene A, Wiebe KD. Land tenure and the adoption of conservation practices. American Journal of Agricultural Economics. 2000;**82**(4):993-1005

[66] Chiputwa B, Langyintuo AS, Wall P. Adoption of conservation agriculture technologies by smallholder farmers in the Shamva District of Zimbabwe: A Tobit application (internet). Available from: https://ageconsearch.umn.edu/record/98851/ [2022.04.20]

[67] Rodríguez-Entrena M, Arriaza M. Adoption of conservation agriculture in olive groves: Evidences from southern Spain. Land Use Policy. 2013;**34**:294-300

[68] Mavunganidze Z, Madakadze IC, Nyamangara J, Mafongoya P. The impact of tillage system and herbicides on weed density, diversity and yield of cotton (*Gossipium hirsutum* L.) and maize (*Zea mays* L.) under the smallholder sector. Crop Protection. 2014;**58**:25-32

[69] Rochecouste JF, Dargusch P, Cameron D, Smith C. An analysis of the socio-economic factors influencing the adoption of conservation agriculture as a climate change mitigation activity in Australian dryland grain production. Agricultural Systems. 2015;**135**:20-30

[70] Uddin MT, Dhar AR, Islam MM. Adoption of conservation agriculture practice in Bangladesh: Impact on crop profitability and productivity. Journal of the Bangladesh Agricultural University. 2016;**14**(1):101-112

[71] Van Hulst FJ, Posthumus H. Understanding (non-) adoption of conservation agriculture in Kenya using the reasoned action approach. Land Use Policy. 2016;**56**:303-314

[72] Ntshangase NL, Muroyiwa B, Sibanda M. Farmers' perceptions and factors influencing the adoption of no-till conservation agriculture by small-scale farmers in Zashuke. KwaZulu-Natal Province. Sustainability. 2018;**10**(2):555

[73] Kolady D, Zhang W, Wang T, Ulrich-Schad J. Spatially mediated peer effects in the adoption of conservation agriculture practices. Journal of Agricultural and Applied Economics. 2021;**53**(1):1-20

[74] Thierfelder C, Wall PC. Rotation in conservation agriculture systems of Zambia: Effects on soil quality and water relations. Experimental Agriculture. 2010;**46**(3):309-325

[75] Giller KE, Witter E, Corbeels M, Tittonell P. Conservation agriculture and smallholder farming in Africa: The heretics' view. Field Crops Research. 2009;**114**(1):23-34

[76] Thierfelder C, Cheesman S, Rusinamhodzi L. A comparative analysis of conservation agriculture systems: Benefits and challenges of rotations and intercropping in Zimbabwe. Field Crops Research. 2012;**137**:237-250

[77] Erenstein O. Cropping systems and crop residue management in the Trans-Gangetic Plains: Issues and challenges for conservation agriculture from village surveys. Agricultural Systems. 2011;**104**(1):54-62

[78] Ahmed T, Ahmad B, Ahmad W. Why do farmers burn rice residue? Examining farmers' choices in Punjab, Pakistan. Land Use Policy. 2015;**47**:448-458

[79] Prasad R, Gangaiah B, Aipe KC. Effect of crop residue management

in a rice–wheat cropping system on growth and yield of crops and on soil fertility. Experimental Agriculture. 1999;**35**(4):427-435

[80] Snapp SS, Rohrbach DD, Simtowe F, Freeman HA. Sustainable soil management options for Malawi: Can smallholder farmers grow more legumes? Agriculture, Ecosystems and Environment. 2002;**91**(1-3):159-174

Section 3

Information Technology Use in Smart Farming

Chapter 3

Information Technology Drivers in Smart Farming Management Systems

Alexy Márta, András Jung and Bálint Molnár

Abstract

The chapter describes the possibilities of collecting digital data on crop and livestock production and their use in "smart farming" systems. Earth drone and spectral mobile mapping technologies can provide plant production-related measures with high temporal and spatial resolution. Remote sensing helps better understand farming patterns and crop management. Improving understanding of the link between remotely sensed data and risk assessment and management in "smart farming" is very important. Controlled-environment agriculture takes advantage of light recipes, related to spectral light-emitting diode (LEDs) and sensors. In livestock farming, analyzing a database of digital data on the environment and livestock individuals can help farmers make decisions better. The heterogeneous digital data from plant and livestock production are collected into a Data Lake. Then the data are processed to transform the data into the proper format for data analytics. Data Warehouse should be integrated into an ERP system that is dedicated to the agricultural environment.

Keywords: smart farming, remote sensing, drone application, precision livestock farming, IoT, data science, ERP, Data Warehouse, Data Lake

1. Introduction

The transition from experience-driven to data-driven decisions in agriculture production is unthinkable without the use of digital tools and solutions. Given the traditional nature of agriculture production and its custom-based practices, this is not a quick process. During the last decades, farming has been forced to implement measures to increase efficiency and productivity at the expense of resilience in the face of climate change and environmental variability. Intensification in agriculture has been causing a serious impact on environmental sustainability. The industry is facing a reduction in the workforce, and consumer demand is growing for more transparent, sustainable, ecological, and high-quality products. Moreover, the new common agricultural policy of the EU aligned with the European Green Deal is focused on environmental protection in rural areas. The use of precision methods – i.e. information and communication technologies (ICT)tools, processes, and methods – in the production of agricultural products is becoming increasingly important in production practice. In the two major sectors of agriculture, crop production and large-scale livestock farming, precision technologies are enabling the creation of "smart farming" systems. In data-driven farming, the expertise of the farmer is becoming more valuable. The results of data analysis that exploits large-scale databases can be incorporated into decision support systems. This data analytics will enable actors in the agricultural sector to rely on accurate, traceable, and credible production results to make decisions that will help them manage cost-effectively and optimize the environmental impact of production.

Digital tools and analytical methods are ready for use in agriculture. It will take time for them to become widespread in farming practices and part of farmers' animal and crop production procedures. The application of data analytics will validate the added value of precision practices and their added value for farmers in real-life farming environments through lots of good practices. This chapter reviews the most relevant precision methods in crop production, large-scale livestock production, and their usability in the decision support process. The latter leads to "smart farming" technology.

2. Precision plant production

The real benefit of proximal and remote sensing is the capability to characterize spatial or field variability that cannot be parameterized more effectively by any other way. This function has great potential for all land-use practices to increase information availability in everyday farming using proximal and remote sensing technologies. Generally, remote sensing performs nondestructive chemical measurements without intrusion into the material, while providing the possibility of a broad spatial overview and high temporal flexibility.

When spatial thematic information is requested for large-scale areas at regular times, satellite remote sensing is often applied in agriculture. Nowadays, many traditional remote sensing tools are available for both large- and small-scale farming entities as well. Spectral imaging and non-imaging sensors are powerful bio- and geochemical data acquisition tools that can play a crucial role in the early detection of crop management risk factors, such as soil nutrition supply, pests, and diseases.

From a practical point of view, those research studies in the paper are considered that are using field spectrometers and/or spectral cameras and attempt to understand the agricultural values, biophysical and biochemical properties, or reactions of cultivated plants. Only outdoor or field-related applications are included in this discussion of the paper.

2.1 Remote sensing data acquisition

The fineness of the spatially distributed data depends on the sensor and platform. There is a technical limitation to the spectral and spatial resolutions of the satellite platforms. This constraint causes that high spectral resolution and high spatial resolution cannot be achieved at the same time from the same satellite altitude. It has technical aspects, one of them being a justifiable signal-to-noise ratio. The signal-to-noise ratio (SNR) compares preferred signal levels to unpreferred ones. It is complex to give an average SNR for a sensor or multispectral data because it depends on wavelengths, radiance levels, and other technical parameters. Generally, non-imaging spectrometers provide higher SNR values compared to imaging ones. Satellites with less than 1-m pixel size have less than 10 broad bands in the spectrum typically, while satellite images with more than 10 spectral bands have larger pixel sizes than 10 m on the ground typically. One way to increase the spatial and spectral resolution is to change the sensor and reduce the altitude of the data capturing. This demand initiated many different forms of terrestrial and near-ground imaging and non-imaging spectroscopy.

The spectral resolution describes the electromagnetic spectrum to sense material properties and characterizes the number and width of the spectral channels available for spectroscopic sampling. The spectral resolution could be also interpreted as the "chemical resolution," since the spectral resolution resolves the apparent spectral material properties and links chemistry to spectroscopy. Accordingly, the higher spectral resolution provides more detailed chemical insights [1].

Temporal resolution is a factor in agricultural remote sensing that controls flexibility and data availability. The periodical returns of satellites are typically not demand-driven, and the airborne campaigns with the high-temporal resolution are very cost-intensive and complex.

Radiometric resolution is a technical term that characterizes the sensitivity of the detector or the wavelength-dependent energy resolving power of a sensor. It is quantified by bits, typically. Accuracy and stability are essential in radiometric calibrations to calculate radiance and/or reflectance that are the derivatives and representative outputs (information carriers) of the remotely sensed data and the primary inputs for further statistical analyses.

In color imaging, three broad bands (blue, green, and red) are used to reproduce real-life object properties in a virtual form the best. The RGB (red, green, and blue) bands are broad spectral channels.

When the number of spectral channels is increased (over 100) and the spectral range is extended (400–1000 nm or more), imaging spectroscopy or hyperspectral imaging is applied.

2.2 Characteristics of data sources

Spatial scales of field phenomena are not absolute and are customized to specific needs and applications. From a global (earth-observing) point of view, scales smaller than 104 km^2 could be referred to as local scales, which are higher by several magnitudes than the common agricultural management scales in Europe. For site-specific observations, further downscaling is needed. For crop management, the variability on the field and subfield scale are of interest, and the variability at distances of 50 m or less is mainly related to management practices [2].

Considering options for the remote sensing application in agriculture is one of the most time-critical. The entire crop sector and production are based on time-critical processes that contain sowing, plant protection, fertilizing, irrigating, and all management decisions.

In spatial downscaling when the measurement height drops down to 100, 10, and 1 m, the temporal, spatial, and spectral resolution can be significantly increased and new demands or application needs such as mobility (e.g. on the fly) and flexibility (e.g. vehicle-based) can be considered.

The temporal resolution affects not only the process accuracy but also the imaging process. Recent developments show that a novel kind of imaging technique (e.g. snapshot spectroscopy) enables high-rate spectral images to generate spectral videos that

Passive remote sensing		Active remote sensing	
Multi-/Hyperspectral	**Thermal**	**Radar**	**Lidar**
Plants: • Leaf pigments • Phenology • Cell and tissue structure • Water content • Biochemical processes and products (lignin and cellulose) • Diseases *Soil:* • Clay minerals • Humus content • TNC • CEC	*Plants:* • Water stress • ET stress • Pathogens • Harvesting • Yield estimation *Soil:* • Moisture • Texture	*Plants:* • Canopy height • Canopy density • Plant height • Canopy structure • Biomass *Soil:* • Soil roughness • Soil moisture • DEM	*Plants:* • 3D plant model • Volumetric parameters • Plant morphology • Canopy structure *Soil:* • High res. DEM • Erosion • Geomorphology

Table 1.
Passive and active remote sensing tools used to characterize plant and soil parameters.

are an obvious advantage in online process monitoring and controlling of agricultural conditions both in field and indoor.

Ref [3] accentuated that characterizing vegetation, soil, or environmental parameters through spectrometers would offer new opportunities. Meantime, many new opportunities for application were found in science and research, primarily. Remote sensing research topics often focused on stress caused by pest or disease incidences, yield and biomass estimation, nutrition deficiencies, drought, frost, etc. Vegetation stress may cause anomalies in the cellular or leaf structure affecting the pigment system or the moisture content in canopy, which could be detected and mapped by optical sensors as can be seen in **Table 1**.

2.3 High-resolution crop remote sensing

Remote sensing of biophysical parameters such as phytomass, leaf area index (LAI) and canopy structure have intensively been analyzed, [4–7]. Behind the biophysical parameters, numerous papers have been devoted to biochemical components such as foliar constituents, chlorophyll a and b, carotenoids, lignin, cellulose, protein, water, and other elements [8, 9]. Many of the studies used high-resolution full-range (FR) spectra (400–2500 nm), because some foliar chemistry components give indications only over 2000 nm (e.g. lignin and cellulose) [10, 11]. Our study focuses on narrowband indications in the range of 400 to 1100 nm. Our study focuses on narrowband indications in the range of 400 to 1100 nm.

Multispectral, satellite remote sensing used broad (50–100 nm) spectral bands initially, which have been narrowed by scientific high-resolution sensors over the last decades [12]. The VNIR (400–1100 nm) spectral range will remain significant in future crop sensor developments as well, but it will be spectrally enhanced likely, to produce high-resolution crop or soil sensors. Band comparisons highlight the best

Wavelength (nm)	Parameter	Indications	References
375	Biochemical	Leaf water content	[13, 14]
466	Biochemical	Leaf chlorophyll	
515	Biochemical	Leaf nitrogen	[14]
520	Biochemical	Pigment content	[15, 16]
525	Biochemical	Leaf nitrogen	[17, 18]
575	Biochemical	Leaf nitrogen	[19, 20]
675	Biochemical	Leaf chlorophyll	[14, 21]
700	biochemical	Nitrogen stress	[14, 22]
720	biochemical	Nitrogen stress	[13, 23, 24]
740	biochemical	Leaf nitrogen	[14, 25]
490	biophysical	Crop yield	[14]
550	biophysical	Biomass	[23, 26]
682	biophysical	Crop yield	[14]
845	biophysical	Biomass	[27]
915	biophysical	Crop yield	[14, 24]
975	biophysical	Leaf moisture	[28]
1100	biophysical	Biomass	[29, 30]

Table 2.
Narrowband sensor wavelengths for measuring crop parameters.

benefits of the narrowband indices such as non-saturating behavior or high sensitivity in vegetation dynamics (e.g. phenology) [13]. The narrowbands can be classified as very narrowbands (1 nm to 15 nm), narrow bands (16 nm to 30 nm), intermediate bands (31 nm to 45 nm), and broadbands (greater than 45 nm) [13]. For future VNIR crop sensor developments, the following spectral narrow bands could be of interest (**Table 2**).

Narrowband studies showed that classification accuracies have increased. Generally, the hyperspectral narrowbands explain about 10–30% greater variability in quantitative biophysical models in comparison with broadband and are not sensible to saturation problems in biophysical estimations [31]. These benefits are to be considered in future high-resolution imaging or non-imaging crop sensors. There are known important parts of the VNIR spectrum: the so-called red-edge region, which is likely becoming increasingly important for novel optical field sensors as well (**Table 2**).

An auspicious tool to detect vegetation conditions is to study the sharp rise of the reflectance curve between 670 and 780 nm. This segment is called the red-edge region. Both the position and the slope of the red-edge point (REP) change due to physiological conditions and can result in a blue- or redshift of the inflection point. The red-edge index is defined as the position of the inflection point of the red-NIR slope of a vegetation reflectance curve. The reliable detection of this index requires high-resolution spectral measurements [27]. The well-known methods are to define red-edge position (REP) [32]. The reflectance curve's numeric derivation and interpolation techniques are also widely used. The REP is correlated strongly with foliar chlorophyll content which is a sensitive indicator for various environmental factors. A comprehensive spectral analysis has been conducted on fruits and other agricultural products in scientific studies [33]. Recent developments in REP offer new perspectives and approaches for spectral mobile mapping services.

2.4 New demands and tendencies

The demand for out-of-the-lab devices initiated the early field spectroscopy and sensor with non-imaging measurements, which originated from laboratory spectroscopy and required respective developments in optics and portable platform techniques. From the beginning, portable or handheld field spectroradiometers were very popular in geology, soil, and vegetation spectroscopy as they provide flexible and rapid field data acquisition [34]. Thus, the spectroscopy in the visible (VIS) and near-infrared (NIR) has been widely used either in the laboratory [35] or for in situ monitoring [36].

There is an apparent gap between integrative point measurements and airborne or even space-borne image data. Field imaging line scanners are less widespread in ground-truthing than portable point spectroradiometers. Non-scanning or snapshot hyperspectral imaging is one possible solution to overcome this limitation of in-field usability [37]. Snapshot hyperspectral imaging enables rapid data acquisition as the entire image with all spectra is captured, at once, within a few milliseconds by a handheld or portable mode [38].

Optical field data acquisition has been reshaped and extended by new platforms in the last few years. This kind of platform liberalization changes our ground-truthing attitudes and fieldwork traditions. Traditionally, field spectroscopy was used to support airborne and space-borne campaigns (**Figure 1**).

The proximal and remote sensing spectral detectors are either imaging or non-imaging sensors. Until recently, light-weighted spectral scanners were not used widely because of technical limitations. One of the first successful fix-wing miniature spectral scanning measurements was achieved by [39]. The light-weighed scanners (< 1–2 kg) mainly work in the spectral range from 400 to 1100 nm. These typically utilize push-broom spectral imaging. Hyperspectral cameras with the scanning principle cannot capture random moving objects. Mobile imaging field spectroscopy requires sensors that are flexible and easy to operate. Non-scanning hyperspectral imaging has been recently introduced for many outdoor applications. Non-scanning spectral imaging is called "snapshot imaging spectroscopy" [37], and it has a different principle from the push- and whiskbroom sensors (**Figure 2**).

A snapshot light-splitting architecture integrated on a sensing sensor chip with appropriate spatial resolution captures the full-frame image with a high spectral (> 100 bands) and radiometric resolution (> 14 bit). The image capturing process benefits from a powerful light collection capacity [37]. For a hyperspectral snapshot camera, in a sunlight situation, the integration time of taking one hyperspectral data cube is about 1 ms. Such a camera can capture more than 10 spectral image data cubes per second, which facilitates hyperspectral video recording. The commercially available snapshot imaging spectrometers record hyperspectral full-frame images with more than 20–100 bands in a spectral range of 400 and 1000 nm.

The snapshot advantage prefers time-critical applications either in the laboratory or in the field. This fact is significant for vegetation studies and crop management, especially because of physiological and phenotypical changes [31]. Knowing more about temporally resolved spectral crop information is of high importance for

A B

Figure 1.
A non-imaging (A) and an imaging spectrometer (B) in field use. Source: A. Jung.

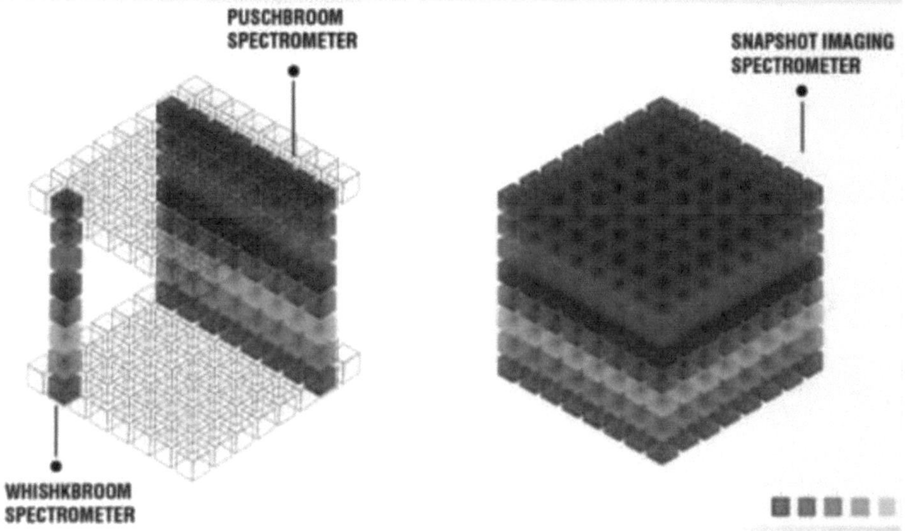

Figure 2.
Working principles of hyperspectral imaging, colors represent different wavelengths. Source: Courtesy of Cubert GmbH, Germany.

agriculture among others because of timely and targeted nutrition supply, and preventive and precision pest control. Beyond the temporal aspect, there is a general and global demand for high-resolution data in agricultural process control. The technical paradigm change in imaging field spectroscopy will enhance the effectiveness and availability of commercial sensors in smart farming.

The real-time image capturing capability of the proximal snapshot imaging spectroscopy is essential for capturing moving objects (i.e. leaves and canopy) or being on a moving platform (i.e. vehicle, unmanned aerial vehicle (UAV), robot, or human being) at high-resolution scales. Smart farming applications could offer individual detection and treatments of species or canopies that are of interest in the viewpoint of economic, environmental, and professional. Reference [40] used a snapshot hyperspectral imaging camera in a farming experiment to study its usability on a UAV platform to monitor crops. This study concluded that the combination of 3D imaging techniques and snapshot hyperspectral imaging enables the precise and accurate monitoring of dynamic crop growth through phenological changes. A multi-temporal crop surface analysis enables the precise monitoring of plant height and plant growth, while hyperspectral analysis derives physiological vegetation parameters like chlorophyll or nitrogen content and others. To monitor crop growth behavior, crop vitality, and crop stress snapshot hyperspectral imaging may be an ideal sensor [40].

3. Precision livestock farming

In all farm animal species, farmers have the same goal: to produce animal products profitably. The scope of inputs is similar: feed, water, livestock, medicine, infrastructure, and human resources. The big difference is that the animal species have different external and internal characteristics, the farms have different technological standards, and the management requirements are different. These factors determine the structure, quality, and use of the resources listed earlier. The digital technologies ought to fit into the operational framework of a given livestock farm and create a well-defined added value for farmers. This is not an easy task under the conditions in which livestock products are produced, the often difficult to predict market environment affecting profitability, and changing social and regulatory factors.

The driving force behind the demand for precision technologies in large-scale livestock farming is the possibility of early detection of diseases primarily [41]. In the case of breeding animals, this offers the opportunity to reduce the additional costs of culling due to the late detection of disease. In these groups of animals, the farmer wants to keep the farm animal in breeding and reproduction as long as possible. The objective of rearing animal products for human consumption (fattening and/or laying flocks) is to produce the animal product of the quality required by the market in the time available, using the optimum quantity of inputs (feed, drinking water, etc.) to reduce the negative environmental impact.

The early detection of diseases and their electronic alerting to the farmer through the analysis of digital data collection tools and the database they produce is just one of the possibilities offered by precision livestock technologies. Achieving this goal also provides the producer with a range of other useful information: the appearance of the disease is indicated by changes in animal behavior. While the digital devices collect data on individuals and send a signal when they change, analysis of the data in the database can reveal several important facts. These include the time spent eating, drinking, and resting, which is typical of all our farm animals. Species specificities can be observed, such as scratching in poultry, wallowing in pigs, or ruminating in cattle. In addition to studying individual behaviors, the observation of social behavior also provides the farmer with useful information about certain farm animals (e.g. fighting or playing patterns in pigs).

These observations have been made by farmers in the past and present without digital tools. The difference is that the use of precision technologies reduces the need for personal presence and allows continuous monitoring of individual and group animal behavior instead of periodic observations [42]. An additional advantage is that farmers can gain practical knowledge and experience not at the time of data collection, but also at a later point in time. The data can be saved and examined in other using methods that were not previously available.

The central issue of precision livestock farming (PLF, smart livestock farming, and smart animal agriculture) is how to increase food production sustainably. While, the farmers should care about animal welfare and reduce the environmental burden. This goal can be achieved by merging data that originated from data acquisition (sensors) and Internet of Things (IoTs). The data transformation along with predictive analytics can be applied by using artificial intelligence (AI) tools. Ref [43, 44] indicated that PLF uses principles and technology of process engineering to manage livestock production through smart sensors to monitor animal growth, milk and egg production, endemic diseases, animal behavior, and components of the microenvironment within the production unit (in Ref [44]). Precision livestock technologies are playing an increasingly important role in response to the factors that hamper the production of animal products. These include reducing the environmental impact of intensive livestock production systems, reducing the cost of inputs by optimizing their quantity (including human labor), and keeping up-to-date knowledge of animal health by studying individual and group animal behavior [42, 45]. A more detailed in-depth analysis of data from the farming environment will help livestock keepers make decisions and indirectly improve their ability to generate income. These technologies consist of digital data collection, the creation of databases from that data, the analysis of the data, and the presentation and visualization of the patterns found in the data. The big difference with traditional methods is that databases containing data on animals or environmental parameters collected by digital means cannot be successfully analyzed using traditional statistical methods. This requires the use of data science methods to identify internal patterns in the data sets that are not possible with traditional analysis methods.

It is important to note that precision technologies, i.e. the use of informatics in the collection and processing of data from livestock, cannot be generalized in practice. There is no universally applicable device or procedure that answers the questions of livestock farmers. The simplest, most reliable, and least costly IT solution must be tailored to the circumstances of the farm. Digital data collection tools can be categorized according to the type of data they collect, and the most appropriate analysis methods should be selected from those already available. This requires knowledge of the specificities of the farm. In practice, these precision solutions will only be widely used if they are validated in on-farm projects.

This section describes the framework for precision farming technologies.

3.1 Data acquisition

Recently, there are several publications on precision livestock farming [46–51]. Several authors have concluded that although the IT solution that are used works well, its practical uptake remains to be seen. In [52], the major limitation of practical applications includes high installation and maintenance cost, difficulties in using the new technologies due to lack of knowledge or skill of farmers, lack of confidence in the manufacturing companies, etc. How can this be changed, and how could this be improved? (**Figure 3**).

The use of precision livestock farming technologies would contribute to consistent objective and regular welfare monitoring of livestock in real time, allowing farmers expeditiously to identify problems and implement preventative measures to avoid

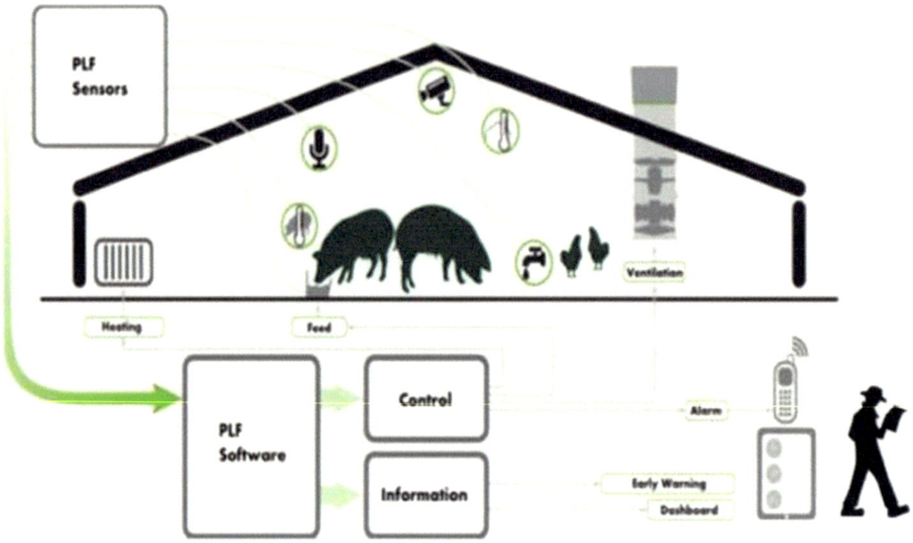

Figure 3.
Farm with precision livestock farming technology [52].

critical failures [53]. In large-scale livestock farming, the data that come into the earlier-mentioned database can be divided into two broad categories: direct (i.e. digital data coming in via IT tools) and indirect (digital data recorded by the farmer). Digital data must be collected from all possible sources to have an accurate database of the production activities of a given livestock holding. This database is necessary to achieve the objective. The so-called analog and historical data that are collected traditionally should also be incorporated into the database.

Another equally important aspect of data collection is the environment-oriented (i.e. data collected directly on the farming environment) and animal-oriented (data collected on livestock individuals) data sets. The use of both animal- and environment-oriented data supports easy and proper monitoring of health, welfare, production, and risks [54]. Farmers have been collecting these data with varying degrees of attention for years. After all, a careful farmer wants to know exactly how much it costs to produce the animal product and where there are points for improvement. However, we can only talk about precision livestock farming if the database also contains digital data on the individuals in the livestock.

3.1.1 Housing system

Table 3 lists the input data that can be measured and analyzed to provide the farmer with information on the farming environment. There are three main types of farming in large-scale livestock production: confined, semi-confined, and free-range. The more enclosed and controlled housing is (poultry and pigs), the easier it is to collect data on the microenvironment of the barn using sensors. These data were collected by farmers for decades, especially in confined housing, because it is the basis for automatic ventilation, cooling, heating, feeding, and water technologies. This approach represents an opportunity for precision livestock farming because farmers do not use this large amount of data for data analysis purposes in most cases. In semi-confined systems, animals are affected by the external environment, and

Sensors	Housing system		
	Closed	Semi-closed	Free-range
Temperature	xxx	x	x
Humidity	xxx	x	x
Air speed	xxx	x	x
Ammonia	xxx	x	—
Carbon dioxide	xxx	x	—
Air pressure	xxx	x	x
Feed level	xxx	x	—
Drinking water flow	xxx	x	—

Table 3.
Data in the farming environment, Alexy M.'s own research.

human influence is minor (dairy cattle). In free-range conditions (beef cattle and pigs), the role of humans is even reduced, with external weather factors fully determining the environmental conditions around the animals. In this type of farming, it is also possible to collect environmental data, e.g. by placing weather stations in the pasture.

The way how animals are kept also has a big influence on the individual data that can be collected about them. **Table 4** lists and groups the most popular digital data collection tools according to their practical application in different housing systems.

Table 4 shows that not all digital devices are suitable for all types of housing systems. The usability of the tools, the cost of acquisition and operation, and the quality and quantity of data that can be collected are of important considerations. The matching and integration of data in different formats into a common database is of particular importance. Ensuring data transmission is one of the most difficult tasks (interoperability) because local data storage must be implemented if there is insufficient bandwidth. These are practical problems that can only be solved, tested, and developed, in the context of pilot experiments.

Device (based on [55])	Housing system		
	Closed	Semi-closed	Free-range
RFID (passive or active)	xxx	xx	xx
Rumen bolus	—	xxx	xx
Walk over weigher	xxx	xx	x
Cameras	xxx	xx	—
UAV	—	x	xxx
GPS	—	x	xxx
Accelerometer	xx	xxx	xx
Pedometer	x	xxx	xx
Microphones	xxx	xx	—

Table 4.
Applicability of PLF tools in different housing systems, Alexy M.'s own research.

3.1.2 Livestock characteristics

A key element of precision livestock technologies is to acquire data on individuals of livestock. The data acquisition requires either placing a digital device on the animal or inside a part of the animal or collecting data on the stock. In the latter case, it is necessary to segment the images during data analysis after data collection, by labeling every animal. In all cases, the behavioral and body characteristics of the animal must be known.

Table 5 presents the digital data collection tools that can be considered for three farm animal species (cattle, pigs, and poultry) and assesses their applicability. The digital tools presented are those included in **Table 4**.

In cattle farming, the milk and meat production directions are indicated separately. The reason for this is the different production purposes of the cattle and the different farming environments. In the case of pigs (although almost 95% of the world's pig population is kept in confined, intensive conditions), free-range farming is also observed (the purpose is the same, here free-range means organic farming and meeting the needs of other target groups of customers). All digital data collection tools can be used in cattle farming. The possibility to collect individual data from poultry flocks that are kept in completely enclosed conditions is hampered by the body structure of the birds. Namely, birds have no external ears to which, e.g. radio-frequency identification (RFID), tags can be attached and their fast growth rate causes problems since devices that can be placed on the neck or limbs cannot be used because the animals' body size changes so rapidly that it may damage the animals' physical integrity. In their case, the use of cameras and microphones is an option. Attempts have been made to use RFID technology on the wingtip, but in practice, this is not a feasible and worthwhile investment.

Device	**Species of livestock**				
	dairy and beef cattle (semi-closed)	**Dairy and beef cattle (free-range)**	**Pig (closed)**	**Pig (free-range)**	**Poultry (deep litter)**
RFID (passive)	xxx	xxx	xxx	xxx	x
RFID (active)	xxx	xxx	—	—	—
Rumen bolus	xxx	xxx	—	—	—
Walk over weigher	x	x	xxx	x	xx
Cameras	xx	—	xxx	x	xxx
UAV	—	xxx	—	xxx	—
GPS	—	xxx	—	xxx	—
Accelerometer	xxx	xxx	—	—	—
Pedometer	xxx	xxx	—	—	—
Microphones	xx	—	xxx	—	xxx

Table 5.
Applicability of digital devices in cattle, pig, and poultry, Alexy M.'s research.

In livestock farming practice, the specificities of the farming environment largely determine the quality of the data that can be collected. A good example of this is our experience in one of our pilot projects, in which we analyzed camera images of poultry flocks to estimate individual weights and detect behavioral anomalies. In this case, the issue was not with the application of the model (artificial neural network), because the results were surprisingly good, but with devices for the data collection and storage. The metal parts of the computer in the enclosure, which collected the camera images, were so corroded after only two fattening cycles that the device was unusable. The camera lenses had dust on them, a spider had woven a web in front of the lens (only whiteness was visible), and the high temperatures and humidity required to capture the day chicks had fogged the lenses. Consequently, we were able to use only some images for analysis. It was not possible to store the images taken by the cameras on a remote computer (Cloud storage), because the site did not have strong reception (edge storage).

3.2 Data storage and preprocessing

At this stage, the skills of data scientists are needed. In the first two steps, the experts sit at the same table, and in this phase, the domain expert leaves the table but stays in the room. Data preparation consists of several steps: cleaning the data, sorting data in different formats, performing anonymization, reconciling data in different tables, etc. Then the most appropriate data analysis model can be selected. One of the cornerstones to select a data model is the type of data collected (image, sound, number, etc.). By analyzing the aggregated database, data scientists can build a data analysis model by identifying the internal patterns in the data series. The data must have the characteristics that allow a credible and correct analysis to be performed. Data science defines at least six important characteristics (6 V's, **Figure 4**), each of which must be present in the data for the database on which the data analysis is based to be usable.

Although this process may seem simple, practical experience shows that it is not. We have experienced this in one of our pilot experiments by ourselves, in which we collected individual data on the daily activity of Mangalica breeding sows kept outdoor. Passive RFID tags were inserted in the ears of 20 sows and readers were installed in an area of the pasture, at the wallowing area. Data were collected on three

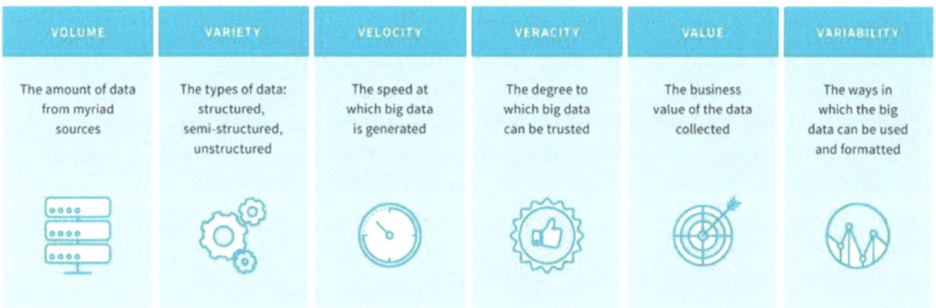

Figure 4.
6 V data characteristics. Source: https://nix-united.com/blog/how-big-data-is-transforming-the-education-process/.

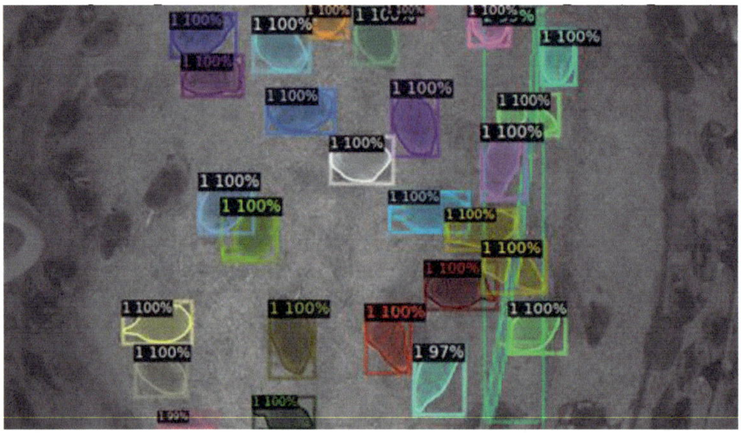

Figure 5.
Poultry is detected by the neural network algorithm on images. Source: Alexy, M.'s own research.

weather parameters at hourly intervals. The database included the time of arrival and departure of the sows (from which the duration of presence can be calculated) and the data of the three weather parameters (temperature, humidity, and air pressure). After cleaning the database, the frequent item sets method from data mining (apriori algorithm) was chosen to obtain information on the daily activity and social behavior of the sows [56]. However, during the evaluation of this model, we found that it was not suitable to evaluate the database over time and to determine the activity trends of sows. After several months of work, a new model was set up and is currently under analysis. Although the business and data understanding was formulated and the database preparation was successful, the use of an inappropriate model could not answer the practical question.

However, the artificial neural network algorithm used to evaluate images of poultry flocks successfully recognized the birds. This is illustrated in **Figure 5**.

3.3 Evaluation and deployment

In this step, the domain expert returns to the table: s/he evaluates the patterns established by the model and determines whether the question asked has been answered correctly. The data scientist may have found correlations that are flawed or irrelevant from an expert's point of view in a livestock domain. If the evaluation shows that the result has added value for the domain expert, the practical application of the solution in the field can begin. This solution will be presented to farmers, then the results are validated whether it has value to farmers. As the practical example aforementioned shows, in the peer review process, it is possible that the desired result is not achieved by applying the wrong model. A livestock professional knows the farming and breeding characteristics of the livestock and understands the complexities involved in producing animal products. They know whether the data science answer to the question asked at the start of the project is appropriate. If so, the results obtained can be applied in practice and disseminated to livestock farmers.

The results of the precision method, validated in a real farming environment, can be successfully integrated into the decision support system of the agricultural enterprise. The outputs of precision methods provide real-time and reliable input information to information systems, which can be used as the basis for complex, strategically important economic decisions. This is presented in the next section.

4. Architecture solution for the problems of smart farming and information supply chain

Recently, agriculture became one of the primary sources of data through the applications of various sensors and IoTs (Internet of Things). In the ecosystem of agriculture, horticulture, and farming, the efficient and effective utilization of data has gained momentum and become an essential issue. Transformation of data collection from the simple substantiated financial data to data that originate from large-scale monitoring and controls of operation led to the requirement of disciplined data analytics. There are data that can be listed as traditional data of company operation and operational data of farming originated from different devices, IoTs. These latter types of data are unstructured and heterogeneous; either we consider their structure or their content. Some pieces of data are accompanied by metadata that may describe the essential information about the content and can be utilized to categorize them and organize them into a data catalog to exploit them for advanced data analytics.

4.1 Agriculture and information management

The development of past decades transformed agriculture and farming regarding digitization and data processing profoundly. Some data sources are available and can be utilized as social media. Web information systems, Internet of Things/sensors, and management information systems of the agricultural enterprises, and electronic images, which were generated by various equipments, have an important role in the assessment of several relevant factors within the production process of farming.

The application of Data Warehouses is apt to structured data that originated from structured databases [57, 58]. The processing of unstructured data by advanced algorithms machine learning and data science, e.g. images, audio, and text, requires cleaned data, a so-called single source of trust that encompasses reliable and trustable data for prediction and prescription. The aforementioned, different types of data can be gathered into an appropriate data architecture that would provide step-by-step transformation (**Figure 6**).

4.2 Data Lake for information management

We suppose that the data originating from the disparate source system are of good quality; if this is not the case, the data preparation process has a built-in procedure for quality improvement and data cleansing. The typical life cycle of data related to farming is showcased in **Figure 6**. The left-hand side of the diagram contains the potential sources of data that can play an essential role in farming, either agriculture, horticulture, livestock management, or any other modern branches.

A hybrid data warehouse that includes a robust Data Warehouse, as a consequence, leads to a data life cycle that begins with transformation, cleansing, and integration. The purpose of constructing a Data Lake is to have day-to-day operations that separate transactional data from data collection devoted to reporting and retrieving. The original goal of the services of Data Warehouse was to query historical data

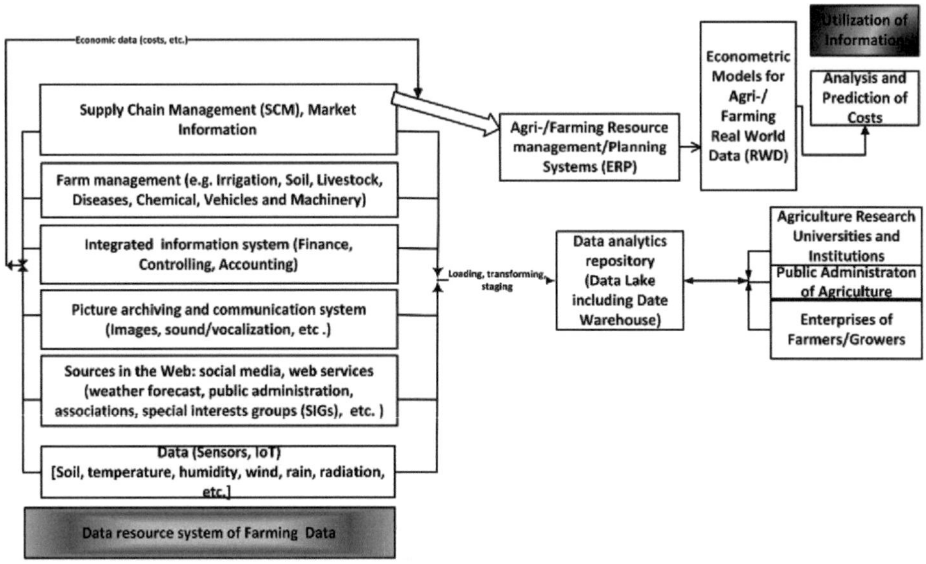

Figure 6.
Data sources and their utilization in smart farming.

and to carry out complex data analysis. In the data preparation stage, the data are cleansed and sieved according to the data structure of the target Data Warehouse, and the major constituents of this general structure are the fact table and the associated dimension tables. The next phase includes activities as follows: data migration, data integration, the transformation of the codes that are used in the succinct description of the data, and conversion to transport data between database management systems. The primary objective of the development of the Data Warehouse was to lay a sound foundation of data analysis in a separate system from the production system to avoid performance problems that may have been caused by complex queries. The ETL (Extract, Transform, Load) process is used to load data into a Data Warehouse. In this step, the general data cleaning and transformation took place, e.g. removal of the last and introductory spaces of data items, removal of redundant zeros, standardization of identifiers/identification numbers, making effective restrictions of data fields; and e.g. conversion of imperial units into metric units of measure or vice versa.

While the aforementioned data manipulation is performed, relationships among data items, tables, and schemas may be violated or lost. Analogously, integrating and combining the data from multiple resources can lead to defects that are transferred into the Data Warehouse. To prevail over the data quality restrictions that were caused by the transformations in the Data Warehouse, the idea of the Data Lake was brought into existence. The idea advertised by the notion of Data Lake is to place the data in its original form in storage areas, i.e. in the Transitive and/or Raw Data Zone after the aggregation phase (**Figure** 7; [59]). Thus, Data Lake can store and obtain data from RDBMS (Relational Data Base Management Systems), semi-structured data (XML, binary XML, JSON, BSON, etc.), and unstructured data (e.g. images), moreover metadata, which are usually displayed in the semi-structured form and characterize the data fed into the Data Lake too. The data entry and preprocessing phase can involve loading, batch, and stream processing for source data while performing the necessary quality checks with the Map-Reduce capability [60]. A vital

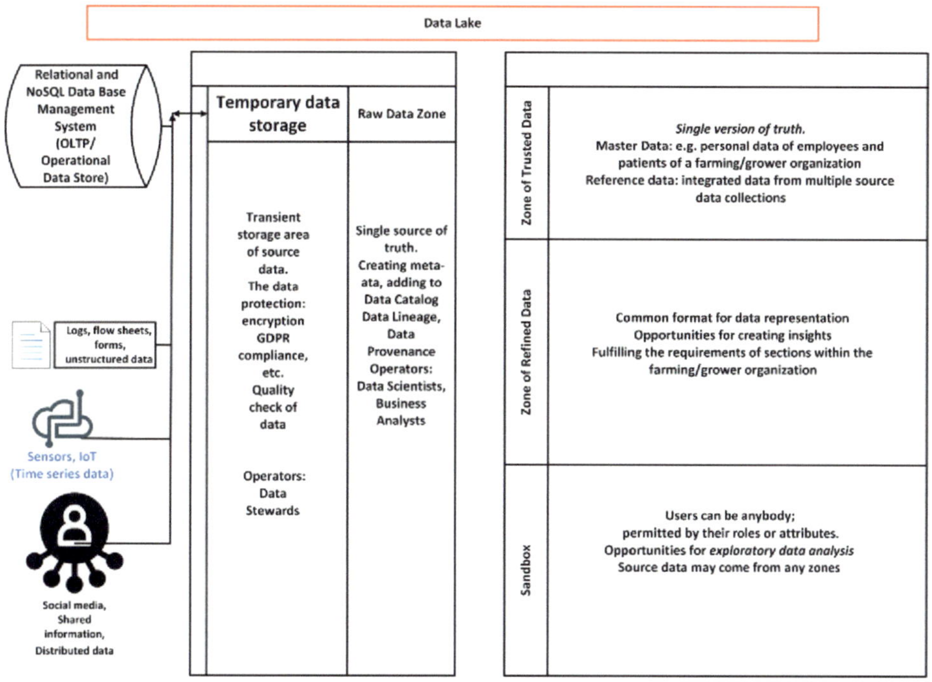

Figure 7.
Data Lake architecture for agriculture.

feature of the Raw Data Zone is that it can be regarded as the "single source of truth" because it retains the data in its original form; however, the data can be anonymized, masked, and tokenized in this zone. Data scientists and business/data analysts can come back to this zone when seeking original connections and relationships among data items that may have become absent during data transformation, conversion, encryption, and encoding. The Trusted Zone implements functions for data processing to ensure quality assurance and to guarantee compliance with standards, data cleansing, and data validation. In this zone, plenty of data alterations take place by the predefined local and global standards, through which the data can be considered "the only version of the truth." This zone encompasses master data and fact data that are registered and tracked by a data catalog that is automatically or semi-automatically populated with metadata. The data in the Refined Zone endure several additional alterations that are designed to make the data usable in data science algorithms. These data manipulations include structuring of the data format, possible detokenization, a quality check of data to meet the yearnings of the algorithms when models of the subject area (e.g. agriculture and smart farming), and data analysis are developed. In this way, procedures for knowledge acquisition through data exploration and analytics can be performed, and comprehension of the data sets can be achieved. Within each zone, user access rights must be rigorously controlled through adequate methods, e.g. role-based access control or other combined ensemble methods that fit the specific environment. In the event of temporary and ad hoc requirements to deviate from the basic settings, attribute-based access rights or any other adequate approaches can be used. For researchers, executives, authorized employees, and other experts who want to conduct exploratory data analytics, sandbox makes it possible to create models for data analytics and discover associations and relationships between attributes, without involving external or internal experts, without other additional costs [61]. The researcher can feed data from any other zone into the sandbox in a controlled way. Interesting results that came into existence can be sent back to the Raw Data Zone for reuse.

4.3 Agriculture and information system architecture

Zachman's framework contains various viewpoints of business stakeholders and a set of models describing the essential facets of overarching information systems. In **Table 6**, the perspectives represent the various layers of the enterprise architecture in the sense of modeling tools, software, and operational infrastructure. The aspects can be perceived as a line of models, in that the lower-level model is a refinement of the upper-level model. The claimed advantage of the Data Lake is that the data extracted from the source systems are transformed before the actual use of the data for analysis. This approach permits more adaptability to requirements than the controlled, structured environment of Data Warehouses.

The purpose of the Data Lake and Data Warehouse dedicated to research within agriculture informatics is to lay the foundation of data analytics workflows. The architecture should support several requirements as follows: (1) achieving processing speed through adjusting configuration parameters; (2) exploitation of the distribution of data among cluster nodes; (3) usage of provenance data; and (4) data placement and scheduling algorithm for input data to prepare the efficient data processing.

The contextual perspective describes the goal to which the system is dedicated. In the case of data analytics workflow in e-agriculture, the objective is to assist the management to formulate the research questions in terms of business processes. At the conceptual level, the disparate models that are devoted to specific research questions and associated with various approaches to data analytics appear. In the logical layer, the specification of the data analytics workflows is exhibited using business process description languages, algorithms of data analytics, and services for data access. The logical level should contain the explicit specification of access rights taken into account of the different regulations, especially GDPR and sector-specific prescriptions. The physical layer contains the technology-specific arrangements and solutions. Besides the actual programs containing algorithms of Data Science, and data preparation activities, the physical level should contain the number of physical-level data processes (executors) that carry out the particular workflow, the number of concurrent tasks, the allocable memory, etc. The implementation and operational layer (the detailed level) depicts the scientific workflows on an infrastructure that contains nodes of data processing that are realized as commodity hardware and the decomposed models into details, respectively (**Table 6**).

4.4 Enterprise resource planning systems, Data Warehouse/Lake for information management in agriculture

In large and small- and medium-sized enterprises (SMEs), the data is transmitted from disparate systems. The essential constituents of the data sources are the Enterprise Resource Planning Systems (ERP), Customer Relationships Management (CRM), and Supply Chain Management (SCM) as the elements of information systems architecture in a company (**Figure 8**). Nowadays, SMEs and even microenterprises apply ERP systems, although open source or open access solutions through Clouds. The requirement is that the data that is produced by ERP systems is to be

Aspects perspectives	What	How	Where	Who	When	Why	
Contextual	Fact, business data/for analysis	Business Service Business Intelligence	Chain of Business Processes, Workflows	The business entity, function	Chain of Business Process, Workflow	Business goal	Scope
Conceptual	Underlying conceptual data model/structured, semi- and unstructured data	Business Intelligence with added value originated	Workflow	Actor, Role	Business Process Model	Business Objective	Enterprise Model
Logical	Class hierarchy, logical data model structured, semi-structured and unstructured data	Service Component Business Intelligence, Data Analytics	Hierarchy of Data Analytics Service Component	User role, service component	BPEL, BPMN, Orchestration	Business Rule	System Model
Physical	Object hierarchy, data model (object store)	Service Component Business Intelligence, Data Analytics	Hierarchy of Data Analytics Service Component	Component, Object	Choreography	Rule Design	Technical Model
Detailed	Data in SQL/NoSQL and other file structures	Service Component Business Intelligence, Data Analytics	Hierarchy of Data Analytics Service Component	Component, Object	Choreography, Security architecture	Rule specification	Components
Functioning Enterprise	Data	Function	Network	Organization	Schedule	Strategy	

Table 6.
A mapping schematically between Zachman architecture and components of Data Lakes [62].

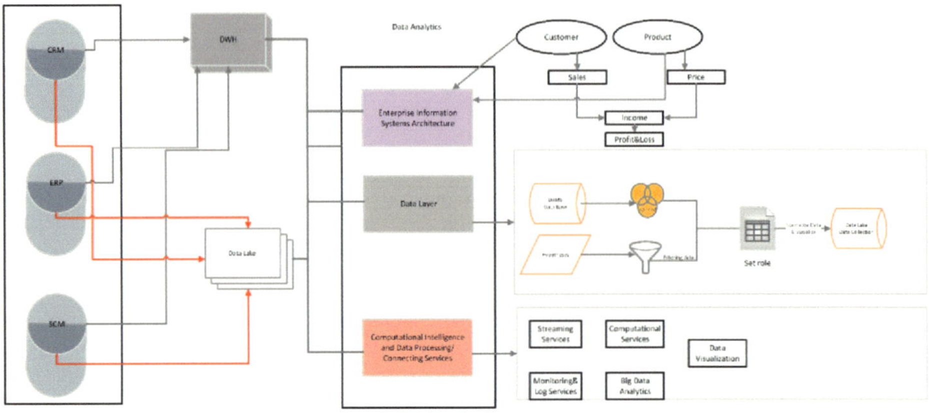

Figure 8.
Enterprise resource planning (ERP) system and Data Warehouse (DWH) and Data Lake architecture for data analytics in agriculture.

stored and archived for later data analytics. Generally, in various industry sectors, Data Warehouses play a significant role in administering data.

The Data Warehouse along with Data Lake in agriculture make it possible to support decision-making, and ground in durable, reliable, and trustable data analytics for the huge size of data both in real time and batch processing (see [63, 64] in other sectors of the economy). The primary role of Data Lake in an enterprise environment is to yield the chance for data integration and reconciliation; the decisive role of Data Warehouse is to provide the opportunity to integrate data in a structured manner and format; moreover, it stores persistently and efficiently the original data in the fact table from the disparate modules of ERP.

Data Analytics and *Business Intelligence* tools can use DW as the major source of structured data and provide insights [65–67]. The structured organization of DW serves as a sound foundation to acquire a holistic perspective of business process performance by senior managers, business workgroups, and data scientists. Generally, the Key Performance Indicators (KPI) can be monitored and tracked. The application of dashboards gives a cleaner, precise, reliable, trustable, and easy-to-access picture of the actual state of the enterprise. The dashboard lays the foundation for effective decision-making. The data feeding or ingestion framework should be built up in the case of Data Lake. In a Data Lake, structured and unstructured data should be handled in a secure environment, considering the data protection requirements. Adequate libraries and workbenches are needed for data analytics tools and machine learning. The data provenance and metadata management can be realized through an advanced machine learning tool set and data catalog in the Data Lake (**Figure 8**). The disciplined and strict access rights should be implemented through sophisticated single sign-on and multi-factor authorization and authentication methods.

Data Quality should be achieved through the application of combined tools in both cases, in DW and Data Lake. The processes of data quality should deal with the erroneous data and pass it to the data quality team for human interaction whether what format can be accepted and inputted.

Integrating Data is an essential function of DW and the operational teams. The goal is to transform the data into the standard structure and format and to create consistent, interpretable, and meaningful data. This process is realized by a data transportation tool that conveys the data from different systems into DW and into the appropriate zone of the Data Lake. This process applies heterogeneous standardization and data cleaning methods that are related to the domain.

5. Conclusion

After showcasing research in agriculture for precision farming in Chapters 2, and 3, we have proposed the application of contemporary, modern information architecture that is capable to process data efficiently and effectively. The focus point is to support decision-making for enterprises including large, SMEs, and even microenterprises. Since there are available solutions in the IT market that can be scaled to the size of the enterprise. Especially, the Cloud-based solutions can be regarded as viable for SMEs and microenterprises.

Acknowledgements

This research was partially supported by the project of "Application Domain Specific Highly Reliable IT Solutions" that has been implemented with the support provided by the National Research, Development, and Innovation Fund of Hungary, financed under the Thematic Excellence Programme TKP2020-NKA-06, TKP2021-NVA-29 (National Challenges Subprogramme) funding scheme.

Conflict of interest

"The authors declare no conflict of interest."

Author details

Alexy Márta, András Jung and Bálint Molnár*
Faculty of Informatics, Eötvös Loránd University, Budapest, Hungary

*Address all correspondence to: molnarba@inf.elte.hu

References

[1] Goetz AFH, Vane G, Solomon JE, et al. Imaging spectrometry for earth remote sensing. Science. 1985;**228**:1147-1153

[2] Adamchuk VI, Ferguson RB, Hergert GW. Soil heterogeneity and crop growth. In: Oerke EC, Gerhards R, Menz G, et al., editors. Precision Crop Protection – The Challenge and Use of Heterogeneity. Dordrecht, Netherlands: Springer; pp. 3-16

[3] Elvidge CD. Visible and near infrared reflectance characteristics of dry plant materials. International Journal of Remote Sensing. 1990;**11**:1775-1795

[4] Mutanga O, Skidmore AK, van Wieren S. Discriminating tropical grass (Cenchrus ciliaris) canopies grown under different nitrogen treatments using spectroradiometry. ISPRS Journal of Photogrammetry and Remote Sensing. 2003;**57**:263-272

[5] Clevers JGPW, van der Heijden GWAM, Verzakov S, et al. Estimating grassland biomass using SVM band shaving of hyperspectral data. Photogrammetric Engineering and Remote Sensing. 2007;**73**:1141-1148

[6] Beeri O, Phillips R, Hendrickson J, et al. Estimating forage quantity and quality using aerial hyperspectral imagery for northern mixed-grass prairie. Remote Sensing of Environment. 2007;**110**:216-225

[7] Schellberg J, Hill MJ, Gerhards R, et al. Precision agriculture on grassland: Applications, perspectives and constraints. European Journal of Agronomy. 2008;**29**:59-71

[8] van der Meer F. Analysis of spectral absorption features in hyperspectral imagery. International Journal of Applied Earth Observation and Geoinformation. 2004;**5**:55-68

[9] Almeida TIR, Filho DS. Principal component analysis applied to feature-oriented band ratios of hyperspectral data: A tool for vegetation studies. International Journal of Remote Sensing. 2004;**25**:5005-5023

[10] Bannari A, Pacheco A, Staenz K, et al. Estimating and mapping crop residues cover on agricultural lands using hyperspectral and IKONOS data. Remote Sensing of Environment. 2006;**104**:447-459

[11] Kokaly RF, Rockwell BW, Haire SL, et al. Characterization of post-fire surface cover, soils, and burn severity at the Cerro Grande fire, New Mexico, using hyperspectral and multispectral remote sensing. Remote Sensing of Environment. 2007;**106**:305-325

[12] Adam E, Mutanga O, Rugege D. Multispectral and hyperspectral remote sensing for identification and mapping of wetland vegetation: A review. Wetlands Ecology and Management. 2010;**18**:281-296

[13] Thenkabail PS, Smith RB, De Pauw E. Hyperspectral vegetation indices and their relationships with agricultural crop characteristics. Remote Sensing of Environment. 2000;**71**:158-182

[14] Thenkabail PS, Enclona EA, Ashton MS, et al. Accuracy assessments of hyperspectral waveband performance for vegetation analysis applications. Remote Sensing of Environment. 2004;**91**:354-376

[15] Wrolstad RE, Durst RW, Lee J. Tracking color and pigment changes in

anthocyanin products. Trends in Food Science and Technology. 2005;**16**:423-428

[16] Gitelson A, Merzlyak MN. Spectral reflectance changes associated with autumn senescence of Aesculus hippocastanum L. and Acer platanoides L. Leaves. Spectral features and relation to chlorophyll estimation. Journal of Plant Physiology. 1994;**143**:286-292

[17] Gitelson AA, Grit Y, Merzlyak MN. Relationships between leaf chlorophyll content and spectral reflectance and algorithms for non-destructive chlorophyll assessment in higher plant leaves. Journal of Plant Physiology. 2003;**160**:271-282

[18] Wessman CA. Evaluation of Canopy biochemistry. In: Hobbs RJ, Mooney HA, editors. Remote Sensing of Biosphere Functioning. New York: Springer; pp. 135-156

[19] Gunasekaran S, Paulsen MR, Shove GC. Optical methods for nondestructive quality evaluation of agricultural and biological materials. Journal of Agricultural Engineering Research. 1985;**32**:209-241

[20] Zhao D, Raja Reddy K, Kakani VG, et al. Corn (Zea mays L.) growth, leaf pigment concentration, photosynthesis and leaf hyperspectral reflectance properties as affected by nitrogen supply. Plant and Soil. 2003;**257**:205-218

[21] Chan JC-W, Paelinckx D. Evaluation of random Forest and Adaboost tree-based ensemble classification and spectral band selection for ecotope mapping using airborne hyperspectral imagery. Remote Sensing of Environment. 2008;**112**:2999-3011

[22] Lichtenthaler HK, Gitelson A, Lang M. Non-destructive determination of chlorophyll content of leaves of a green and an Aurea mutant of tobacco by reflectance measurements. Journal of Plant Physiology. 1996;**148**:483-493

[23] Sims DA, Gamon JA. Relationships between leaf pigment content and spectral reflectance across a wide range of species, leaf structures and developmental stages. Remote Sensing of Environment. 2002;**81**:337-354

[24] Penuelas J, Filella I, Lloret P, et al. Reflectance assessment of mite effects on apple trees. International Journal of Remote Sensing. 1995;**16**:2727-2733

[25] Merzlyak MN, Gitelson AA, Chivkunova OB, et al. Non-destructive optical detection of pigment changes during leaf senescence and fruit ripening. Physiologia Plantarum. 1999;**106**:135-141

[26] Yang F, Li J, Gan X, et al. Assessing nutritional status of Festuca arundinacea by monitoring photosynthetic pigments from hyperspectral data. Computers and Electronics in Agriculture. 2010;**70**:52-59

[27] Filella I, Penuelas J. The red edge position and shape as indicators of plant chlorophyll content, biomass and hydric status. International Journal of Remote Sensing. 1994;**15**:1459-1470

[28] Danson FM, Steven MD, Malthus TJ, et al. High-spectral resolution data for determining leaf water content. International Journal of Remote Sensing. 1992;**13**:461-470

[29] Ustin SL, Roberts DA, Gamon JA, et al. Using imaging spectroscopy to study ecosystem processes and properties. Bioscience. 2004;**54**:523

[30] Abdel-Rahman EM, Ahmed FB, van den Berg M. Estimation of sugarcane leaf nitrogen concentration using in situ spectroscopy. International Journal

of Applied Earth Observation and Geoinformation. 2010;**12**:S52-S57

[31] Zarco-Tejada PJ, González-Dugo V, Berni JAJ. Fluorescence, temperature and narrow-band indices acquired from a UAV platform for water stress detection using a micro-hyperspectral imager and a thermal camera. Remote Sensing of Environment. 2012;**117**:322-337

[32] Smith KL, Steven MD, Colls JJ. Use of hyperspectral derivative ratios in the red-edge region to identify plant stress responses to gas leaks. Remote Sensing of Environment. 2004;**92**:207-217

[33] Zude M, Herold B, Roger J-M, et al. Non-destructive tests on the prediction of apple fruit flesh firmness and soluble solids content on tree and in shelf life. Journal of Food Engineering. 2006;**77**:254-260

[34] Milton EJ, Schaepman ME, Anderson K, et al. Progress in field spectroscopy. Remote Sensing of Environment. 2009;**113**:S92-S109

[35] Ben-Dor E, Banin A. Near-infrared analysis as a rapid method to simultaneously evaluate several soil properties. Soil Science Society of America Journal. 1995;**59**:364-372

[36] Stevens A, van Wesemael B, Bartholomeus H, et al. Laboratory, field and airborne spectroscopy for monitoring organic carbon content in agricultural soils. Geoderma. 2008;**144**:395-404

[37] Hagen N. Snapshot advantage: A review of the light collection improvement for parallel high-dimensional measurement systems. Optical Engineering. 2012;**51**:111702

[38] Jung A, Vohland M, Thiele-Bruhn S. Use of a portable camera for proximal soil sensing with hyperspectral image data. Remote Sensing. 2015;**7**:11434-11448

[39] Zarco-Tejada PJ, Guillén-Climent ML, Hernández-Clemente R, et al. Estimating leaf carotenoid content in vineyards using high resolution hyperspectral imagery acquired from an unmanned aerial vehicle (UAV). Agricultural and Forest Meteorology. 2013;**171-172**:281-294

[40] Bareth G, Aasen H, Bendig J, et al. Leichte und UAV-getragene hyperspektrale, bildgebende Kameras zur Beobachtung von landwirtschaftlichen Pflanzenbeständen: spektraler Vergleich mit einem tragbaren Feldspektrometer. Photogramm-Fernerkund-Geoinformation. 2015;**2015**:69-79

[41] Halachmi I, Guarino M. Editorial: Precision livestock farming: A 'per animal'approach using advanced monitoring technologies. Animal. 2015;**10**:1482-1483

[42] Berckmans D. Precision livestock farming technologies for welfare management in intensive livestock systems: -EN-Precision livestock farming technologies for welfare management in intensive livestock systems -FR-Les technologies de l'élevage de précision appliquées à la gestion du bien-être animal dans les systèmes d'élevage intensif -ES-Tecnologías de ganadería de precisión para la gestión del bienestar en sistemas de ganadería intensiva. Revised Science Tech OIE. 2014;**33**:189-196

[43] Wathes CM, Kristensen HH, Aerts J-M, et al. Is precision livestock farming an engineer's daydream or nightmare, an animal's friend or foe, and a farmer's panacea or pitfall? Computers and Electronics in Agriculture. 2008;**64**:2-10

[44] Tedeschi LO, Greenwood PL, Halachmi I. Advancements in sensor technology and decision support intelligent tools to assist smart livestock farming. Journal of Animal Science. 2021;**99**:skab038

[45] Schillings J, Bennett R, Rose DC. Exploring the potential of precision livestock farming technologies to help address farm animal welfare. Frontiers in Animal Science. 2021;**2**:639678

[46] Chung Y, Lee J, Oh S, et al. Automatic detection of Cow's Oestrus in audio surveillance system. Asian-Australasian Journal of Animal Sciences. 2013;**26**:1030-1037

[47] Vranken E, Berckmans D. Precision livestock farming for pigs. Animal Frontiers. 2017;**7**:32-37

[48] Tullo E, Finzi A, Guarino M. Review: Environmental impact of livestock farming and precision livestock farming as a mitigation strategy. Science Total Environment. 2019;**650**:2751-2760

[49] van der Burg S, Bogaardt M-J, Wolfert S. Ethics of smart farming: Current questions and directions for responsible innovation towards the future. NJAS Wagening Journal of Life Science. 2019;**90-91**:1-10

[50] Maes DGD, Dewulf J, Piñeiro C, et al. A critical reflection on intensive pork production with an emphasis on animal health and welfare. Journal of Animal Science. 2020;**98**:S15-S26

[51] Pandey S, Kalwa U, Kong T, et al. Behavioral monitoring tool for pig farmers: Ear tag sensors, machine intelligence, and technology adoption roadmap. Animals. 2021;**11**:2665

[52] Guarino M, Norton T, Berckmans D, et al. A blueprint for developing and applying precision livestock farming tools: A key output of the EU-PLF project. Animal Frontiers. 2017;**7**:12-17

[53] Neethirajan S, Kemp B. Digital livestock farming. Sensor and Bio-Sensor Research. 2021;**32**:100408

[54] Piñeiro C, Morales J, Rodríguez M, et al. Big (pig) data and the internet of the swine things: A new paradigm in the industry. Animal Frontiers. 2019;**9**:6-15

[55] Aquilani C, Confessore A, Bozzi R, et al. Review: Precision livestock farming technologies in pasture-based livestock systems. Animal. 2022;**16**:100429

[56] Alexy M, Horváth T. Tracing the local breeds in an outdoor system – A Hungarian example with Mangalica pig breed. In: Tracing the Domestic Pig [Working Title]. London: IntechOpen; 2022. DOI: 10.5772/intechopen.101615

[57] Kimball R. The Kimball Group Reader : Relentlessly Practical Tools for Data Warehousing and Business Intelligence. Indianapolis, IN: Wiley; 2010

[58] Golfarelli M, Rizzi S. Data Warehouse Design: Modern Principles and Methodologies. Bologna, Italia: McGraw-Hill, Inc; 2009

[59] LaPlante A, Sharma B. Architecting Data Lakes. O'Reilly Media Sebastopol, 2014.

[60] Duggal R, Khatri SK, Shukla B. Improving patient matching: Single patient view for clinical decision support using big data analytics. In: 2015 4th International Conference on Reliability, Infocom Technologies and Optimization (ICRITO) (Trends and Future Directions). 2015. pp. 1-6

[61] Myatt GJ. Making Sense of Data: A Practical Guide to Exploratory Data Analysis and Data Mining. Hoboken, New Jersey: John Wiley & Sons; 2007

[62] Zachman JA. A framework for information systems architecture. IBM Systems Journal. 1987;**26**:276-292

[63] Molnár B, Pisoni G, Tarcsi Á. Data Lakes for insurance industry: Exploring challenges and opportunities for customer behaviour analytics. Risk Assessment, and Industry Adoption. 2020

[64] Pisoni G, Molnár B, Tarcsi Á. Data science for finance: Best-suited methods and enterprise architectures. Applied System and Innovation. 2021;**4**:69-69

[65] SAP Business Objects Business Intelligence suite. 2020. Available from: https://www.sap.com/products/bi-platform.html [Accessed: May 29, 2022]

[66] Make better decisions, faster with Tableau. 2020. Available from: https://www.tableau.com/products [Accessed: May 29, 2022]

[67] Go from data to insight to action with Power BI Desktop. 2015. Available from: https://powerbi.microsoft.com/en-us/desktop/ [Accessed: May 29, 2022]

Section 4

Smart Farming Approach

Chapter 4

Perspective Chapter: Physiological Breeding Approach for Sustainable Smart Farming

Raja Shankar, Panamanna Mahadevan Govindakrishnan, Shashi Rawat and Joseph Sherly

Abstract

Smart farming is referred as managing farm efficiently using information and communication techniques to increase the quantity and quality of the product. The basic and fundamental concept of smart farming in agriculture is to exploit yield determinants efficiently so as to attain genotype x environment interaction zero by introgression of trait of interest demanded by the environments. Accordingly, the physiological breeding approach coupled with mega environment concept could be a sustainable smart farming, which could be exploited to fulfill the future food demand. This chapter is conceptualized with scientific information available on potato under India such as low land tropic scenario.

Keywords: trait driven breeding, mega-environment, yield determinants, sustainable smart farming, physiological breeding

1. Introduction

The projection shows that feeding the world population of 9.1 billion by 2050 would require raising overall food production by 70%. Considering the future food demand, the trait-driven breeding approach is essential to exploit natural resources to obtain high yield/quality potential of a region in smart manner. The currently developed varieties under conventional approaches seldom express its high genetic potential in every mega environments due to lack of identification of mega environments and its demanding trait of interest for improved yield. Presently, the improved varieties are developed in an environment and tested them at across the environments and selected the best-performed locations for recommendation. Due to which, testing of varieties developed at across locations frequently exhibits low to moderate yield potential as it lacks the specific traits of interest for high potential. Hence, physiological/ideotype breeding for mega environments could result with high potential varieties. Thus, this chapter deals about potato as test crop, present growing scenario and future strategies for yield enhancement considering trait-driven breeding approach for mega-environments concept.

Potato is the fourth most important food crop in the world and is also the most consumed food crop grown in >125 countries. It is top ranked food crop, which supplies 80% edible dry matter. According to Statistics, 52% of crop area lies in the temperate region (Europe), 34% in Asia, and 14% in Africa. Asian and African countries had low productivity as against Europe where it is quite high. However, the estimated potential yield of potato ranged between 40 and 140 t/ha under optimal growing environments [1] depending upon the length of growing season and temperature. India is the second largest producer of potato in the world by producing 53.11 million tons from 12.247 million hectares during 2020–2021. This has been possible due to the adaptation of the crop to different agro ecological conditions in India. Due to which, it has now been spread from hills and plateau (>800 M msl) to the plains (<300 M msl.). The plains presently account for more than 80% of the Indian potato acreage (**Figure 1**), and the hills and plateaus account about 15%. Under plains, about 90% of total production is restricted in subtropical plains, mainly Indo-Gangetic belt (76%), 6% in the hills, and about 4% in plateau region of peninsular India.

Considering the future potato demand of the world, Bradshaw [2] clearly predicted that the leading production primarily happened not only through area expansion in potato, but also accompanied by increased productivity. These trends cannot be continued as such, as the demand gap should be primarily reduced through by increase in productivity as new land will not be so readily available for area expansion under potato in future. Although, India contributes significantly in terms of production, the yield potential realized among the growing zone was very wider due to ecological diversity. An encouraging finding on the similar line, the modern potatoes had high harvest index (0.80) and tuber fresh-weight yields of 120 t/ha achieved in Western Australia prevailed with a long growing season, absence of pests and pathogens with adequate inputs of water and fertilizers [3]. Although these yields are not achieved in practice presently, it supports that the long growing season (hills and plateau) would contribute significantly to meet the future demand of potato in India provided the genetic potential of genotypes is enhanced. Secondly, attaining the yield potential of temperate countries by the tropics and sub-tropics is not easily possible due to shorter growing season with lower efficiency per unit of intercepted radiation due to prevalence of high temperature in Subtropics. However, the countries with long crop season (England) had achieved yield potential from 22 t to 45 t/ha despite the area of cultivation reduced to half during the period from 1993 to 2006 (British Potato Council Statistics) due to the intervention of superior yielding varieties. Hence, the limited area having long growing season (hills and plateaus) and larger area having

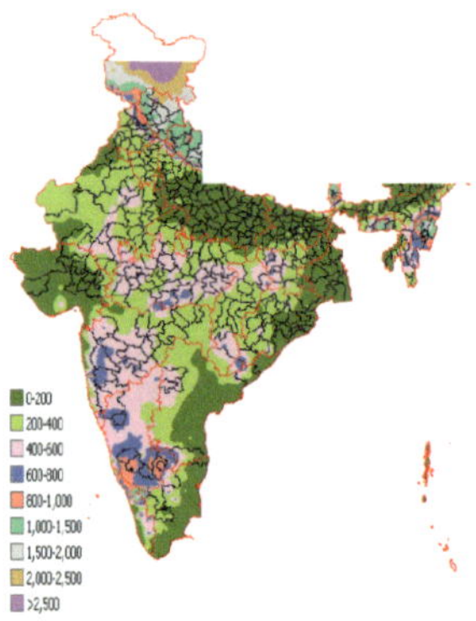

Figure 1.
Potato growing districts in India and their elevation.

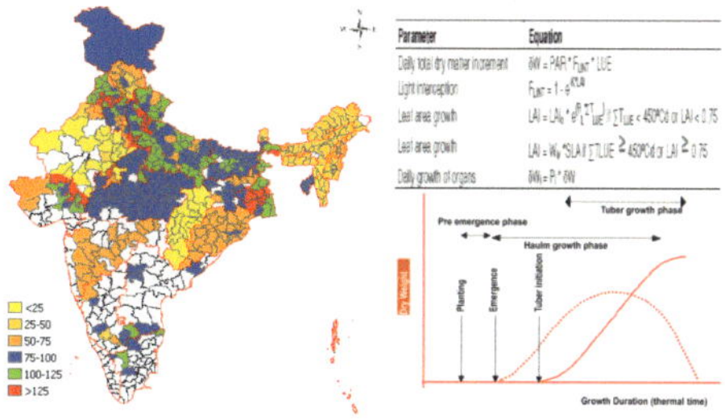

Figure 2.
Potato relative yield index estimates across India.

short growing season (plains) should be exploited systematically using smart farming for productivity enhancement to meet the future demand of potato in India.

Physiologically, the potato productivity is influenced by environments that include the internal (genetic) and external factors (climatic, edaphic, biotic, physiographic, and also socioeconomic factors). Although, potatoes have been adapted to warm climate and grown successfully in tropics, tuber yields remain lesser [4] due to an unfavorable allocation of assimilates within the plant. Yield is the outcome of the product of incident radiation, radiation use efficiency, and harvest index.

Haverkort and Harris [5] reported that temperatures above 23°C favor allocation of dry matter to the foliage at the cost of tuber growth. The pictorial diagram given below indicates the dissecting the potato yield into different components (daily total dry matter increment, light interception, leaf area growth, and daily growth of organ vary from growth phase and thermal region of growth phase. Hence, the genotypes utilizing the available resources efficiently are better yielder. Rawat et al. [6] reported that potato relative yield index estimated across locations in India had ranged from less than 25% to more than 125% indicating wider variability (**Figure 2**). Harvest index is a measure of the proportion assimilates partitioned into harvested organs. Plant physiologist and breeders adapted harvest index as a selection index for breeding high-yielding cultivars; however, Gawronska et al. [7] opined that high harvest index may not necessarily correlate with high yield. A cultivar is able to produce high rate of carbon assimilates and maintain active growth later in the season, thereby giving high yield in spite of the harvest index value, which implied that in addition to climate and edaphic factors (environment), the genotype interaction for the maximum resource use efficiency is also very much essential.

2. Understanding the climate diversity

The above-ground environments (temperature, humidity, rainfall, photoperiod, and solar radiation) and below-ground environments (soil type, nutrient and moisture, pH, etc.) determine the yield potential. In order to exploit the growing environment, there is need for testing of genotypes at multilocations (MLT) in the hills, plateaus, and plains across the country to recommend the best variety to its most adapted domain. The figure of climatic normals estimated had the mean maximum temperature of growing period of 90 days ranged from 21.1°C to 32.6°C at 25 locations, where the minimum temperature was recorded between 8.0°C and 20.9°C. This diversity can be exploited to evaluate the hybrids for their performance under different stress levels. Considering the above fact, the process of photosynthesis is very sensitive to high temperature where the optimum temperature for photosynthesis in potato is reported to be about

20°C, and at an increment of 5°C above, the optimum is expected to decrease photosynthetic rate by 25% [8]. The optimum canopy net photosynthetic rates in potato have been reported at 24°C, while the maximum biomass accumulation has been reported between 18 and 20°C [9, 10]. Accordingly, the 25 centers can be classified according to mean temperature of the growing season, where in Pantnagar, Srinagar, and Jalandhar, the mean temperature of the growing season is expected to be quite low, and hence, the growth and development of the crop would be slow, and hence, at these centers, the crop could experience low temperature stress. This is also reflected in the thermal time accumulated for 90 days <1600 at these centers. The optimum temperatures of 18–20°C would be prevalent in Kanpur, Shillong, Shimla, Patna, Hissar, Dholi, Chinwara, Kota, and Modipuram. Hence, these centers are ideal for evaluating genotypes for their yield potential under optimal temperature condition. This is also reflected in the accumulated GDD at these locations, which ranged from 1600 to 1800 degree days. Mild stress is characterized with mean temperature of 20–22°C, which prevailed in Gwalior, Kalyani, Jorhat, and Raipur, while high stress (with mean temperature > 22°C) recorded in Bhubanewswar, Pune, Deesa, and Darwad. This finding clearly proved that these four centers are the target locations for evaluating genotypic performance against heat stress conditions. Additionally, the mean temperature of these four locations was reflected high as corresponding to high-growing degree days accumulated (>2000^{0}Cd). A wider variability in mean night temperature of the growing season ranges from 11.3°C (Pantnagar) to 22.9°C (Dharwad). Similar variability also exists for incident solar radiation. The yielding ability of a genotype is an outcome between the reactions of the genotype interaction with different agro-ecological conditions. Knowledge on G x E is very important for choosing varieties to realize high yields and stable production [11]. Small genotypic reactions due to changed environmental condition are desirable in agricultural production [11], since the genotypes with a minimum yield variation at varied environments are considered stable [12, 13]. It has also been proven from the multilocational trials (MLTs) that the genotypes did not follow the similar trend at across the locations tested due to their interaction pressure against the environment pressure.

In physiological sense, the life cycle of potato can be divided into two stages (growth stage and tuber stage). The growth stage begins immediately after breaking the dormancy of tuber results with sprouting. However, in agronomical sense, the growth stage starts after planting of tuber in the field and continues till tuber mature. The tuber stage begins from tuber attaining maturity to subsequent its planting as seed. According to Pushkarnath [14], there is an extended time required for

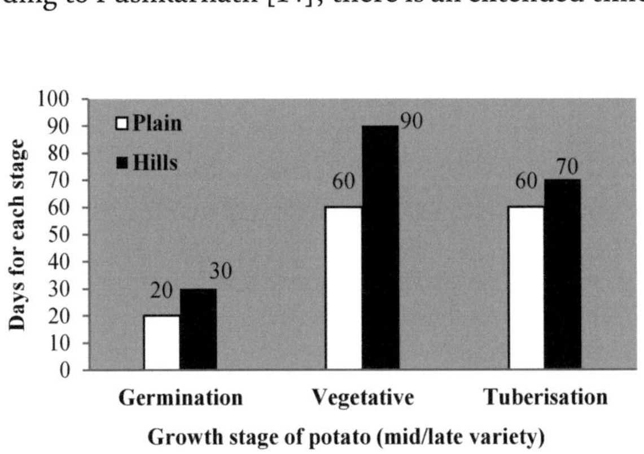

Figure 3.
Growth stage specific duration differential in mid/late genotypes of potato at hills and plains.

the same genotype of potato grown under hills for germination (10 days), vegetative (30 days), and tuberization (10 days) as compared with growing in plains (**Figure 3**). So, the same genotype grown under hills required 40–50 days extra period to complete its life cycle than that grown in plains. This extra period encounters with all other environment components' pressure to result in better yield. Hence, clear dissecting of environmental components contribution of a location needs to be determined to extrapolate in the climate similarity areas to recommend the genotype.

3. Identification of mega environments

A huge environmental variation and varied genotypic behavior necessitated to use specialized technique for Genotype x Environment and its interactions. A common way of screening genotypic reaction to environmental factor is "multi-environment trials" (METs). In MET, a number of genotypes while evaluated at a number of geographical locations for a number years encounter different pattern of stresses, which predicts the response of genotype for future growing environments as their stress tolerance level differs from each other. In order to study a Genotype x Environment interaction, different kinds of trials are being adopted and regional yield trials with a kind of network of experiments having a set of cultivars performances assessed to make a genotype recommendations [15]. The test location within the target region should adequately represent [16], having major cropping areas and farm practices in order to reflect the variation in climate, soil, biotic, and crop management factors. The test site should not be lower than 6–7, and the trials time should be about 2–3 years to distinguish repeatable from non-repeatable GxE interaction effects.

A recent concept for identifying target domain of genotype with minimum field trials is the mega environment concept [17]. The basic principle behind mega environment concept is the interaction pattern between the genotype and environments. Griffith et al. [18] proposed four types of interaction between genotype and environments, where additive (A), divergence (B), and convergence (C) interaction showed a parallel or non-parallel relation between genotype and environment due to which one genotype (either G1 or G2) showed superior yield consistently at both environments with slight differences in its yield level. Whereas the crossing-over (D) interaction is the most useful model, where the reaction norms are non-parallel, and there is a strong Genotype x Environment Interaction (GEI), and the relative difference between the yield level of G1 and G2 varies among the environments, and the low-yielding genotype in E1 may be superior-yielding genotype in E2. Hence, a mega environment is defined as a group of locations that consistently share the same best cultivar(s) [19], and mega environment could divide the target environments into *meaningful* mega environments to deploy different cultivars for different mega environments for utilization of positive GEI and to avoid negative GEI [20]. Due to which, the cultivar-location interaction pattern can be repeatable across the years. Based on the adaptability of cultivar selection, the target environments are classified into three groups, such as single simple ME, multiple ME, and single complex ME, which have unique feature as given in **Table 1**.

In order to assess the crossing-over interaction, both AMMI and GGE biplots are the data visualization tool, which graphically displays GEI in a two-way table [17] to identify mega environment and candidate genotypes. According to Kroonenberg [21], the GGE biplot analysis is based on environment-centered PCA, while AMMI analysis is double-centered PCA where AMMI stands for the additive main effect and multiplicative interaction [22], and GGE biplot stands for genotype main effect plus GEI [23].

Sl.No	Parameters	Single simple mega environment	Multiple mega environment	Single complex mega environment
1	Number of mega environment in a location	Single	Many	Single
2	Objective	To select single best variety	To select specifically adapted genotype for each mega environment	To select generally adapted genotype across the whole region
3	Genotype and Environment interaction	No crossover interaction	Crossing-over interaction exists	Crossing-over interaction exists
4	Repeatability of crossover interaction over the year	Non-repeatable	Repeatable	Non-repeatable
5	Need of multi-year/ location testing	Not required	Required	Required

Table 1.
Classification of mega environments with unique feature.

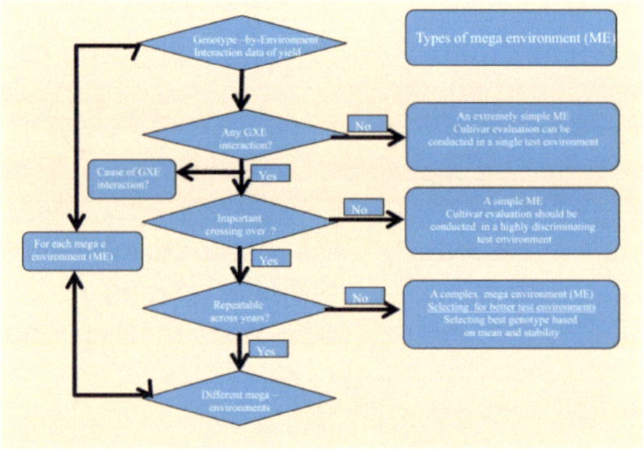

Figure 4.
Types of Mega environments and its genotypes selection approach.

Hence, the GGE biplot has many visual interpretations of any crossover GEI [17]. GGE biplot technique explains the relation among the environments, discriminating ability of an environment, stability of genotypes, and comparison of genotypes with ideal genotypes. Hence, these factors explore the relationships between genotypes and environments. The genotypes with more similarity to each other are very closer to each other positioned in the same plot than genotypes are less similar. The same is true for environments too. Genotypes x Environments that are alike tend to cluster together and the angle between environmental axes is related to the correlation between the environments (**Figure 4**).

4. Environmental determinants of yield

Yield of edible or economical part of a plant is an ultimate goal of any improvement, which relied heavily on modifying the phenotype of crops and finding its genetic basis for its adaptation. Hence, a very successful intervention is to modify the phenological pattern of crop so as to avoid stress [24], to minimize the occurrence of stress through the modified traits (root system) that permits accessing water from deeper in the soil during drought conditions [25], and to match the transpiration rate and evaporative demand [26]. Because, improving the genetic potential of crop for yield depends on introducing the right adaptive traits into broadly adapted, high-yielding agronomic backgrounds. Mega environment and physiological breeding would exploit the environment as well as genetic potential of a plant better than conventional breeding. The following three environment determinants influence yield, and manipulating them according to the need is the priority of the hour.

4.1 Incident solar radiation

The tissue temperature is altered by the rate of radiant energy absorbed by plants, which consequently changes the metabolic process due to energy exchange and results in transpiration. Secondly, the visible fraction of the incident solar radiation can be utilized in the synthesis of reduced carbon compounds (photosynthesis). Thirdly, the energy of specific wave lengths in the solar spectrum can be used by plants as cues for growth strategies. As an example, the red: far red ratio that can influence plant form and dry matter distribution among plant components and the diurnal duration (photoperiod) of incident solar radiation, which can influence the rate of development.

$$PAR_{abs} = PAR_0\left(1 - \mathrm{e}^{-\mathrm{k}LAI}\right) \tag{1}$$

Photosynthetically active radiation incident is equal to half of the solar radiation [27].

4.2 Radiation use efficiency

Under optimum growing conditions, the biomass productivity of different species is defined by the amount of solar radiation intercepted by the green foliage and the efficiency with which such intercepted radiation is converted into plant dry matter. Hence, for most crop species, which grow in the absence of biotic and abiotic stress, the amount of dry matter produced is almost linearly related to the amount photosynthetically active radiation (PAR) intercepted by its green leaf area [28, 29]. Hence, the regression line of cumulative crop biomass versus accumulated intercepted PAR is defined as radiation-use efficiency (RUE) [30]. RUE is referred to above-ground dry matter produced per unit of intercepted PAR (gMJ^{-1}).

$$RUE_{\mathrm{p}} = \frac{TDM}{PAR_0 * \mathrm{F_p}} \tag{2}$$

4.3 Harvest index

Harvest index (HI) is defined as the ratio of harvested edible produced to total shoot dry matter. In potato, cultivars grown in temperate climates with favorable agrometeorological conditions reach the high HI values (0.75–0.85), for the final crop [1, 31–33]. According to reports, it is possible to reach HI of about 90% (HI = 0.90) biomass distribution to tubers [1]. However, the yield has almost doubled in cereal crops as the result of genetic manipulation by plant breeding, despite no change in the rate of photosynthesis per unit leaf area. Alternately, the total photosynthesis has increased as a result of an increase in leaf area, daily duration of photosynthesis, or leaf area duration. Hence, there are opportunities to alter crop duration and the time of crop development to match it to better to radiation, temperature, and vapor pressure during crop growth and to increase the rate of development of early leaf area to achieve rapid canopy closure. Hence, identification of selectable traits in a breeding program to improve crop photosynthesis is helpful to consider the components of biomass production. Assuming there is no water limitation, biomass production is the product of the solar radiation over the duration of the crop period (Q), corrected for the amount intercepted by the crop canopy (I), and the conversion of this chemical energy (E) into plant dry matter. Hence, to increase the duration of crop photosynthesis, there should be an increase in total solar radiation received.

5. Exploitation of yield dveterminants for greater productivity

5.1 Enhanced duration of crop photosynthesis

Considering potato under subtropical low lands, enhancing crop duration is not possible as it has very short crop duration due to short winters. Alternately, hills and plateaus have longer suitable crop duration; however, there is lack of late maturing genotypes. Extending crop duration is the simplest genetic way to increase total photosynthesis, crop biomass, and yield because longer crop duration simply increases the solar radiation interception (Q). The duration of growth that can not only be manipulated to increase biomass and yield, but also its timing should be taken care off. Full light interception should be achieved by the time daily solar radiation is at its maximum when the manipulation for crop phenology better matches to the periods of high radiation with critical growth stages. A high photo-thermal quotient is favorable since high radiation results in increased photosynthesis, whereas low temperature results in slower development during the critical period of high radiation, adjusting phenology by genetically manipulating development time so that the tuber development stage coincides with a high photo-thermal quotient. However, the period of higher solar radiation with water is a limiting factor, then maximizing growth when conditions are cool and vapor pressure deficit is low will increase water use efficiency and biomass production. This can be applicable for *kharif* potato genotypes development program.

5.2 Enhanced interception of solar radiation and rate of photosynthesis through leaf traits

Early growth of leaf area: In cereals, larger and bold-seeded genotypes give probably larger seedlings and thereby more vigorous and larger plants. Also, larger leafier young plants establish more quickly. In potato too, larger tuber size may influence vigorous seedling. Selection for both these traits would result in more competitive crop with greater early radiation interception and faster crop growth rates for which leaf dimension also should be prioritized [34]. Numerous opportunities exist to increase light interception genetically during the early development period of crops; however, lodging is the phenomenon in potato that where the solar radiation interception is hindered. Hence, *non-lodging* would help more in exploiting solar radiation and higher yield needs attention in the breeding program.

Specific leaf area: SLA is the ratio of leaf area to leaf weight, and the speed of emergence is also found to be important [35]. The lodging resistance genes allow the cultivars to grow with more fertilizer under adequate rainfall or irrigation water (higher inputs) results in substantially greater crop yields. In wheat, the non-lodging genes (*Rht1* and *Rht2*) increased the harvest index without reducing above-ground biomass, thereby resulting in greater yields. Similarly, the slower leaf area growth associated with these dwarfing genes comes about from a delay in emergence [36] and from the reduced cell size in wheat [37]. Although there is a higher SLA, a lower assimilation rate is resulted due to reduction in the amount of photosynthetic machinery per unit leaf area. Hence, the increase in leaf area compensating for the reduction in photosynthesis through greater light interception early in crop development is advocated.

Leaf Area Index: Longer the green leaf area enhances the photosynthesis duration. In Potato, the leaf area index starts declining after 40–50 days of emergence, which is the stage when sink strength would be more due to tuber formation and its development, which require more assimilation for higher yields. LAI is another important means to increase total crop photosynthesis and hence, biomass production through increased or extended light interception. Longer duration of leaf photosynthetic activity has contributed to increased yield in most of our major crops. It is difficult to separate the effects of genetics to increase photosynthetic duration due to better nutrition, genetic resistance to foliar diseases such as late blight, and differences among genotypes in nitrogen allocation to tuber or increased demand for photosynthates due to sink strength and its interaction on late blight incidence. A little change in the rate of leaf photosynthesis per unit area accompanies the substantial genetic increase in yields of wheat, rice, sorghum, soybean, sugarcane, cotton, brassica, and sunflower. However, the higher yields have been associated with a decline in the rate of photosynthesis per unit leaf area relative to that of their progenitors. At early growth stage, a high SLA resulted with a higher NAR on a per unit leaf weight basis as against the later growth stages at which the SLA declined, thereby providing an important means for increased photosynthesis per unit leaf area canopy closure [38].

6. Genetic approach for exploitation of mega environments

6.1 Conventional breeding approach

In traditional breeding, exchange of genes between two plants to produce offspring that have desired traits is done, which also takes a long time to achieve desired results and frequently, traits of interest do not exist in any related species. The development of new cultivars is acyclical process, whereby each cycle consists of three major phases [39–41] such as (i) generating genetic variability through making crosses, inducing mutation, introducing exotic germplasm, and using genetic engineering techniques; (ii) selection and testing of identified superior recombinants; and (iii) release, distribution, and adoption of new cultivars: the yield testing in multi-environment trials (METs). However, the efficiency of a traditional breeding program is less commonly measured in terms of selection gain (or response to selection) obtained in a particular cycle or as the average of a number of cycles, and as benefit/cost ratio, which has been used mostly by economists in impact studies.

6.2 Physiological breeding approach

Although, mega environment concept identifies the homogeneous environments, the yield realization depends on the traits of cultivars, which would enable full utilization of the environment. Donald [42] stated that too little attention had been paid to the basic process governing dry matter production and its transformation into economic yield. Therefore, Donald [43] developed a breeding approach for small grain cereals based on the design of model plants (or) ideotype rather than by the traditional breeding. This approach gained momentum with the finding of a linear relationship between accumulated biomass and intercepted radiation as proposed by Monteith [28, 29]. He introduced the concept of radiation use efficiency (RUE), which represents the efficiency of conversion of radiation energy into biomass. Based on this principle, yield is determined as the product of intercepted radiation, radiation use efficiency, and partitioning of total assimilates. In addition to identifying a genotype, the methodology of physiological breeding aims to identify critical traits of the genotype associated with three drivers of yield and breeding to introgress them into genotypes (Ideotype approach). Fischer [44] has proposed a black box approach, which begins with evaluating a set of genotypes in the presence of a known constraint of interest (e.g., moisture stress/heat stress) and concurrently measuring putatively useful traits in the same genotypes so that the critical traits that have a high genetic correlation with economic performance are identified. This approach has been adopted for identifying genetic variability in water use efficiency [45] and its role in adaptation to drought. Contrarily, the ideotype approach aimed at the physiological understanding to predict what features of an ideal genotype for the target environments might be. This approach is extended by use of crop growth models to simulate the expected consequences of traits in target environments [46, 47]. Rasmusson [48] developed an ideotype for barley consisting of 14 traits, in which two were phenological, and the remaining were morphological. The physiological breeding approach is being adopted in many crops [49] including potato. In potato, Haverkort and Kooman [50] developed a method to define an ideotype of potato for a given situation, which involved accounting for yield defining, limiting, and reducing factors on crop growth parameters and their influence on the length of the growth cycle. In the Indian context, the effect of growing season temperature is expected to reduce the length of the crop cycle from 90 days (Jalandhar) in the North-West Indo-Gangetic Plains to 73 days (Bhubaneswar) in the Eastern Plains with corresponding yield reduction from 5.07 t/ha (Jalandhar) to 3.7 t/ha (Bhubaneswar) on dry weight basis due to higher mean seasonal temperatures prevailed in the Eastern plains. Hence, to develop well-adapted varieties, the variability in drivers of yield in different genotypes has to be studied in the light of the protocol developed by Haverkort and Kooman [50] to develop adapted varieties to the widely different environments prevailing in India. Hence, the physiological understanding for assisting plant breeding has to be prioritized [51].

A critical issue in establishing and conducting a breeding program is determined by resources and choice of breeding strategies to breed and select varieties to meet stakeholder needs. The core breeding as mentioned in the diagram indicated that the generalized sequence of activities, which involves recurrent process of selection and crossing, which aimed to achieve an improvement in overall breeding value of the breeding population with time by increasing the frequency of favorable alleles. The criteria for selection of genotypes vary with the phase of genetic improvement in the breeding program (**Figure 5**).

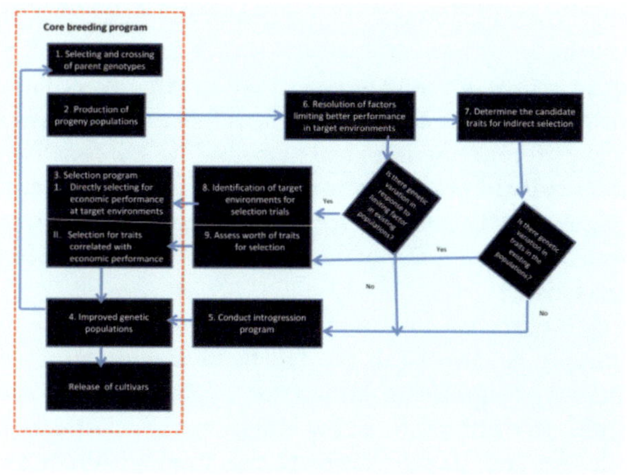

Figure 5.
Core vs. physiological breeding pattern for special traits to be introgressed for specific domain. Schematic diagram of core breeding (left) and physiological understanding (right) [51].

The most common, crossing and selection are done to directly develop varieties for commercial release by selecting both parents and their progeny based on either direct estimates of overall economic worth in targeted environments (e.g., based on measurements of economic value per hectare in multi-environment trials) or on measurements of characters correlated with this economic worthiness (selection program @ box 3).

However, in a more strategic phase of genetic improvement (conduct introgression program @ box 5), is to select genotype for use as parents on the basis of a specific character (e.g., resistance to a disease, tolerance to a stress), which is to be introgressed [52] into locally adapted breeding stocks. The donor genotype(s) of the character being introduced may be inferior from an overall agronomic point of view; however, several generations of crossing and selection are required to combine adequate expression of the introduced trait(s) with a satisfactory agronomic background. Hence, the products of an introgression program found to be desirable to enter the "core" program where selection is based on estimates of total economic performance.

Correlated response to selection for yield gain: Usually a particular environment is chosen for various reasons (e.g., an experiment station) because of its convenience or representing particular feature of the general region being targeted and also often selected for particular characters (e.g., disease resistance). Gain in the character of economic interest (e.g., mean yield across the targeted environments) from selection for any other character (e.g., yield measured in particular environment, disease resistance rating, a physiological trait) may be considered using the following formula for indirect selection [53]:

$$\mathrm{CR} = ih_x r_g, \sigma_{gy}, \tag{3}$$

where CR is the correlated response of character y, the character of economic interest, from selection for the character x, the character on which selection is based; i is the standardized selection differential (as hx is the square root of the heritability for character x, $r \sim$ is the genetic correlation between character y and character x, and σ_{gy}, is the genetic standard deviation for character y.

In all physiological research aimed at an outcome involving genetic improvement, some knowledge of the environmental factors and/or associated physiological processes limiting performance of relevant genotypes in targeted environments (resolution of factor limiting better performance in the target environment @ box 6) such as moisture stress at different phenological periods, low soil fertility, partitioning of assimilate to economically important organs, and radiation interception during reproductive growth (**Figure 4**). The use of crop growth models to quantitatively assess constraints has been advocated [47], which may have value where highly variable seasonal conditions exist.

Determining genetic variation for major limiting factors: There are two factors involved in determining genetic variation. The factor having large effect on the economic performance of significant proportion of genetic population under evaluation at single or multi-environments. There is significant genetic variation in response to the limiting factor for genetic populations being used within or available to the breeding program.

Common procedure to demonstrate genetic variation for a particular constraint is to compare performance of relevant genotypes in one or more environments where performance is markedly affected by that factor, with performance in other "control" environments where the constraint is removed [54]. An alternate approach to identify similarities (or dissimilarities) among trial environments for environmental factors that may cause variation in response among genotypes (e.g., moisture stress at particular phenological stages, soil characteristics, presence of diseases). These similarities are related to relationships among environments for trends in genotype responses. Pattern analysis methodology (e.g., principal component analysis, cluster analysis) may be a useful tool. Identification of environmental factors that vary among selection environments in a similar way as patterns of genotypic response will lead to testable hypotheses about the role of those environmental factors in causing GE interactions.

A selection program in core breeding, GxE interaction is significant, as it typically is, gain from selection will be affected by the choice of environments for selection trials [55]. Ideally, the ranking of genotypes based on results from selection trials should have a high correlation with what would theoretically be obtained if the genotypes were evaluated across all the targeted environments [56]. Therefore, identification of key constraints for which genetic variation in response exists can provide guidance for choosing types of environments for selection trials (identification of target environments for selection trails @box 8). Where GE interaction is insignificant, any small random sample of environments for selection trials may be near optimal, and gains through more deliberate choice may be difficult to obtain. However, where the GE interaction is large, identification of casual factors is expected and one or a couple of factors account for much variation. In a small number of selection trials, which are carefully sited or managed to ensure appropriate levels of the relevant factors, may lead to, on average, significantly greater gains than if a similar number of random environments were used.

Identification of major physiological or environmental factors limiting performance of existing cultivars, coupled with background physiological understanding of plant response to those limiting factors, has led to many suggestions of physiological characters that may be selected by breeders [24, 43, 48]. It has been proposed that first is to use physiological traits as indirect selection criteria in core breeding programs (determine candidate traits for indirect selection @ boxes 7 @ boxes 7 and Asses worth of trait for selection @ boxes 9). The second is to provide objectives of focused introgression programs (determine candidate traits for indirect selection @ boxes 7 and conduct introgression program @ box 5). Both approaches rely firstly on identifying a trait that affects performance for the character of direct interest (e.g., yield) in targeted environments using black box and ideotype approach [44]. The ideotype model was first advocated by Donald [43], which aimed at the physiological understanding to predict the features of an ideal genotype in the target environments. This approach is extended by use of crop growth models to simulate the expected consequences of traits in target environments [46, 47].

Both the ideotype and black box approaches have some advantages and disadvantages. The ideotype approach (including the use of crop growth models) suffers a major limitation in that it does not, by itself, consider the key genetic parameters such as the traits identified in an ideotype may have little genetic variation or have adverse genetic correlations with other useful traits. The ideoype approach therefore may not identify those traits that could be changed for which gains via breeding may be easiest. Whereas the black box approach partly addresses such issues and it suffers in that the results are relevant only to the genetic population represented by those genotypes being examined. In most physiological studies, the genotypic range is very limited and not representative of breeding populations of relevance to breeding programs. However, the ideotype approach may identify traits of value from outside the germplasm pool being used by breeder [48].

To assess the value of a trait for indirect selection in core breeding programs, the parameters of heritability and genetic correlation with economic characters need to be estimated. These parameters together determine the gain from indirect selection and allow comparison with other selection method. Where GE interaction for the character of economic interest is large, care needs to be taken in determining genetic correlations with putative traits. As per example, osmotic adjustment and early flowering have both been shown to improve yield performance in dry environments [57]. For such traits, it is appropriated to determine the genetic correlation between these characters and economic performance in the full range of target environments. A negative correlation with performance in some environments (e.g., in high rainfall or irrigated environments) may impact on how the traits should be used in selection. In some situations, a negative correlation with yield in some environments and a positive correlation with yield in other environments may justify breeding for specific adaptation if the different types of environments are repeatable and if the effect of the character on economic performance is large.

Physiological breeding approach has been showing increasing impact in countries such as Australia [58] as well as in CYMMYT's maize and wheat breeding programs. Selection for reduced anthesis-silking interval in tropical maize has significantly boosted yields under drought conditions. In wheat, a new generation of drought adapted lines developed by combining stress adaptive traits (reduce radiation load-wax, pigment composition, leaf angle, and rolling) have been released as part of CIMMYT's 27th Semi-Arid Wheat Screening Nursery in 2010. The efficient screening has allowed elite genetic resources to be identified in large collections of landraces and using them in strategic crossing [59]. Fine-tuning of phenotyping approaches has also facilitated gene discovery through developing experimental populations in which phenology is controlled, as well as through implementation of rapid screening (e.g., measuring canopy temperature) that permit precision phenotyping of large numbers of genotypes within a time frame that does not confound measurement with environmental fluxes [60].

7. Identification of mega environments for sustainable productivity

The genotype productivity is significantly reduced when the genotype is unable to use the full capacity of favorable environmental conditions. All the varieties are selected for specific agro-ecological conditions and only in such conditions they utilize their maximum genetic potential (with the use of optimal agrotechnology). Jovovic et al. [61] reported the results of 3-year study of productivity for the five leading potato varieties in Montenegro: Riviera and Tresor (early), Kennebec (medium-early), Aladin, and Agria (medium-late). The highest yield of all investigated varieties and localities was measured at variety Agria (30.0 t ha-1), while the lowest at Riviera (24.6 t $ha^{-}1$). In this investigation, Agria variety was favorable for yield of potato tuber. Similarly, the productivity variation in each state belonging to different ecological zone showed significant variation, which indicates that there is scope to increase productivity in the poor yielding zone by introducing high-yielding varieties (AICRIP, 2008), the average productivity differential of the country is wider (18.33 t/ha as against the lowest yield 4.21 t/ha).

Hence, to maximize yield throughout crops in heterogeneous growing regions, despite differences in cultivar rankings from place to place due to GEI, frequently it is necessary to subdivide a growing region into several relatively homogeneous mega environments and to breed and target adapted genotypes for each mega environment. However, the identifying mega environments should fulfill four criteria such as flexibility in handling yield trials with various designs, focus on that fraction of the total variation that is relevant for identifying mega environments, duality in giving integrated information on both genotypes and environments, and relevance for the primary objective of showing which genotypes win where. The AMMI model meets these criteria effectively when the usual biplots are supplemented with several new types of graphs designed to address questions about mega environments. The preliminary results indicated that a small and workable number of mega environments often suffice to exploit interactions and increase yields. Accordingly, the identification of mega environments for the following factors would enhance productivity in any crop, in particular the case study crop potato demands the following criteria to enhance productivity in low land tropics.

7.1 Mega environments for high yield

The spatial and temporal changes in potato cultivation in India have extended its cultivation to stressed conditions or situations, which are not fully conducive for high production efficiency. Hence, there is need to address the sustainability of potato production systems under climate change scenario in view of the expected changes in CO_2 and temperature levels. The increased productivity has to be brought by deployment of varieties matching to its most suitable environment using the concept of mega environments. For which, identification of mega environment for high yield ware, seed, and processing potatoes is essential. Because, the major production constraint is seed material production. Presently, Himachal seed potatoes harvested during October–November cannot be used as seed for plains in the same year, and it can be used only in the next year as the freshly harvested tuber needs at least 60 days to break its dormancy by the time the planting season in plains is over (November). Hence, the mega environment for supplying potato seed tubers during August–September needs to be identified so as to plant them in November in plains. Further, using larger seed tuber for planting consumes lot more investment for seed cost to the farmers, due to which the seed tubers are cut into pieces for planting, which causes disease incidence to growing tuber resulting with higher mortality. Hence, the mega environments in which smaller tubers are produced at higher number and can be used for seed tubers so as to avoid cutting of tubers.

7.2 Screening of germplasm for pest and diseases

Being a tropical country, there has been wider variation on climate; yield level, pest and disease prevalence (11–61%) [62]. Improving yield potential combined with biotic and abiotic tolerance is necessitated. In the last five decades, the collection of genetic resources has not acclimatized to varied ecological zone through recurrent mass selection. Study showed that aphid prevalence in the plains is found severe during December–March, during the peak period of potato cultivation at plains in India. This causes severe virus transmission and poor yield in potato. Hence, the seed production program is restricted only at higher hills (>2000 m) where the potato infestation by aphid is very low due to low temperature, but demands long maturity genotypes. For plains, early maturing genotypes are required to raise due to short winter period, which needs to identify an aphid-free mega environment in plains for rising short duration seed potato and supply to plains and vice versa.

7.3 Developing homomorphic floral genotype for high TPS production

The genetic base of *Tuberosum* has also been widened through creation of broad based populations of long day adapted *S.tuberosum* subsp. *andeigena*, and long day adapted *S.phureja/S.stenotomum* [63]. The cultivated potato (tetraploid) is self-incompatible, has very narrow genetic base causing crop susceptible to several diseases. Multiplication through seed is carrier of virus as it happens through tuber multiplication. Hence, identification of mega environment for self-compatible lines selection where homomorphic flower with having early bulking ensures self seeds under natural condition. Although, true potato seed has added advantage of non-transfer of virus to the next generation, low mass, and easy transport and saving of 30–35% input cost in using tubers as planting material. Developing short-duration TPS families, early foliage maturity, and early tuber bulking would be ideal for TPS production. According to Gopal [64], out of 133 accessions tested for possibility of producing hybrid seed and open pollinated seeds, only five accessions, namely CP1330, Hibinskyrany, I-1039, yy6, and yy12, were promising for production of open pollinated seeds as these genotypes produced more than five berries/plants with more than 200 seeds/berries. The accessions showed profuse flowering genotypes (female) used in hybrid seed production found and were late foliage maturing and presumed to be unsuitable for plains for TPS production. In India, *S.tuberosum* sub sp. *andigena* was mainly used as parent to generate genetic variation for short day condition. However, it also tends to impart late maturity in the progeny due to which 11 out of 21 cultivars released for subtropical condition are medium to late maturing type. Though the ssp. *andigena* has been used to some extent in the Indian potato breeding, all the improved varieties produced true potato seeds in only hills and majority of them do not set flowers at plains under natural condition.

7.4 Heat and drought-tolerant genotypes

All over the world, using wild species in potato crop improvement is relatively slow and conservative [65] might be due to differences in ploidy, sterility barriers, etc. Sterility and tetraploidy in conjunction with a high level of heterozygosity greatly reduce the ability to use traditional methods of breeding, specifically hybridization and selection, for potato improvement. Joseph et al. [66] screened the same set of potato genotypes at three locations varying in altitudes (Kufri, Modipuram, and Jalandar), temperate and subtropical plains of India. The magnitude of genetic parameters was found similar at Modipuram and Jalandar and higher than the location Kufri, which indicates either one location can be used for screening of germplasm or separate screening is needed for higher altitude location. Further, the tuber yield did not have significant correlation with plant height or number of leaves per plant in subtropical plains (Modipuram and Jaladhar) as against the location Kufri, where the higher number of shoots had positive correlation with tuber yield (r = 0.55) at hill conditions [66]. Raj et al. [67] evaluated heat-tolerant genotypes and varieties of potato against leaf hopper and mite at one location Modhipuram and found that genotype HT/92–621, the most efficient against hopper burn. Patel et al. [68] evaluated eight potato varieties assessing their suitability for very early and late harvest for premium market price in Gujarat condition, the results revealed that there was significant effect for genotype and environment interaction for both harvest. K. Phukraj had wide adaptability, as against K. Badsha, which was ideal stable genotype for late harvest, and K. Badsha and K. Jawahar were adapted to favorable environments only. For long time, there was few photo-insensitive genotypes of potato recommended for commercial cultivation, large part of peninsular region of India unfits for potato cultivation. Minhas et al. [69] recommended K. Surya for early maturing, heat-tolerant potato for North Western plains and Peninsular India. The tuberization ability of K. Surya was found higher at 24°C over 27°C [69] due to high rate of photosynthesis [70] under multi-location trials.

7.5 Processing genotypes

Considering processing genotypes, the relationship between specific gravity and dry matter content of potato is vital traits, and the relationship has been found to vary with the variety, location, season, and the year of cultivation [71]. Kumar et al. [70] reported that under water weight basis for conversion of specific gravity, dry matter content, and starch content in potato grown in North Indian plains. The dry matter content and the amount and kind of sugar in particular cultivars are inherited characteristic and influenced by cultural or environments [72]. The influence of short day condition on dry matter distribution of tuber at Modhipuram was found the highest distribution at stem end as compared with pith region in K. Chipsona and Atlantic with the highest dry matter content (17%). Patel et al. [73] reported varietal variation for yield, French fry grade tuber at three locations in Gujarat, and it was due to genotype and environment interaction and none of the genotypes performed better at three locations. Pandey et al. [74] screened processing quality potato genotypes under different hilly locations and found that K. Himasona recorded the highest tuber yield and processing grade tubers with field resistance to late blight at all locations in India.

7.6 Nutrient use-efficient genotypes

In potato, the classification of early, mid, and late maturing genotypes characterized to subtropical and temperate climate suited respectively based on the early dry matter accumulation for early harvest of tubers. Singh et al. [75] assessed the performance of early, medium, and late maturing genotypes of potato under subtropical plains, the results revealed that the dry matter accumulation in early genotype (K.Asoka) was at par with medium (K. Giriraj, K. Sherpa and K. Phukraj) and late maturing genotype (K.Sinduri) at 28 days after planting. At 39 days, the early maturing genotype at par performed with medium (K.Sherpa and K.Phukraj) and late maturing genotype (K.Sindhuri), where K. Giriraj, K. Lauvkar performed below par with early maturing type. Up to 71 days growth, the trend was similar as that of 39 days except that the genotype K. Giriraj pickup its dry matter accumulation and at par with early maturing genotype (K.Ashoka) and the late maturing genotype exhibited exorbitant level of accumulation over early and medium maturity group. From 81 to 91 days growth, the accumulation of dry matter in early maturing genotypes was found lower than medium maturing (except K.Lauvkar) and late maturing genotypes. At 115 days, there was no accumulation of total dry matter in the early and medium maturing genotypes except late maturing in which dry matter accumulation process continued. This indicates that dry matter accumulation is genotype-specific but not in early and medium maturing group. According to environment, few genotypes of medium maturity can also suit for early maturing group, for which mega environment concept facilitates screening of genotypes for different maturity group at different environments. The genotype-specific variation in dry matter accumulation has been reported for rain-fed condition, heat stress area [66], frost stress areas, disease stress, and nutrient use efficiency [76], which vary from location to location of potato growing regions, as the uptake of soil nutrients or water is done by contact between the root cells with soil nutrients. Because the nutrient uptake efficiency of plant increases when the root grows to the area where nutrient or water is available, i.e., root interception. The contact also occurs due to the movement of nutrients from the soil to root surface for which the soil must have more nutrient content. The movement of nutrients from soil to roots occurs in both mass flow and diffusion method [77]. In the former, the nutrients flow from the soil toward the roots in solution form, whereas in the latter, the nutrients flow to the adjacent areas where its concentration is lower. At low as well as optimum soil nutrient levels, the soil diffusion supplies much higher ion quantities from soil to roots than mass flow. Hence, diffusion is therefore fundamental importance for the availability of nutrients of nutrients to plants growing in soil [78]. Hence, not only soil water and nutrients but also biotic stress and heat-related stress cause changes in plant system and which alter the uptake of nutrients for dry matter accumulation in potato, ultimately deciding the yield potential. Hence, mega environment for crossover interaction with different maturity of potato is to be identified.

8. Strategy for exploitation of physiological breeding in Indian potato

In addition to identification of mega environments, development of improved varieties with high yield potential is determined by the canopy cover, radiations use efficiency, and partitioning ratio of genotypes [5]. Under unstressed condition, the radiation use efficiency remains constant in time both for total [5] and tuber dry matter production in potato. Even the two high-yielding genotypes follow different strategy among them to result into high tuber yield [31], as the potato cultivar "Moemoe" had better yield primarily through an increase in tuber numbers as against "Moonlight," which had produced more tubers and increased tuber weight in response to irrigation. The estimated mean water use efficiency (WUE) ranged from 13.1 g/lit (Agria) to 6.4 g/lit (Tutaekuri), which equates to 7.6 and 15.6 lit, respectively, to produce a 100 g tuber. This value is much lower than an estimate of the mean global virtual water content of 25lit for a 100 g potato tuber [79]. Hence, the physiological traits of the cultivar Agria would be an index for screening genotypes for drought environments and for WUE efficiency enhancement in Indian potatoes. Ninety-eight germplasm lines were screened for water use efficiency and transpiration rate and found wider variation water use efficiency in Indian potato [69].

Leaf character: The amount of total radiation intercepted by green active foliage depends on the amount of solar radiation and on the proportion that intercepted (based on leaf area index, leaf angle, and scattering nature of leaf). Hence, the need of selection for several traits into one genotype is must to attain higher yield. Considering critical traits, optimum leaf temperature for photosynthesis in potato has been reported to range between 16°C and 25°C, and the increase in leaf temperature beyond this level drastically reduces the photosynthetic rate [80]. There has been wider variation in leaf number per plant, which ranges from 30.3 (Granola) to 93.3 (Agria) [81], and progeny selection for greater leaf photosynthetic rate and leaf shape [82] indicated that they are heritable in nature. Larger surface area in leaves has been associated with increasing light absorption and shown reduced or almost zero expression of lobes and edges in potato [83]. The presence of complex edges and lobes in larger leaves will enable them to disperse absorbed heat very rapidly, which emphasize that the presence of lobes and edges on margin of leaves of potato would help in heat release mechanism at heat stress area. Similarly, waxy surface is usually observed in younger leaves functioning to prevent (or) minimize the transpiration rate of the plant. On the contrary, the smaller leaf area under stress in a range of species is reported to be associated with greater vein density that may contribute to increased abiotic stress tolerance [84]. In several species, the vein density is correlated with hydraulic conductivity of water and maximum photosynthetic rate in leaves [85]. Broader leaves are stabilized by a set of major parallel veins, narrow and mid-sized leaves are stabilized by a central midrib with rectangular branching laterals [86]. The venation systems of leaves are independent of size and which also reveal the water transport system in the plant.

Leaf area index: A high value of LAI has been suggested to indicate a denser or healthier crop canopy; while a low value represents sparse and or drier canopy. The LAI ranged from 1.40 (Dakchip) to 6.60 (Pungo) and have been observed among the potato genotypes. Atlantic, Chipbelle, and DTO-33 showed no decline in their LAI up to 73 days after planting under abiotic stress environment, indicating its tolerance ability [87, 88]. Crop growth rate increases with leaf area index, particularly at early growth stages. Because the relative increases in the interception of photosynthetically active radiation (IPAR) are largest when leaf is small [89]. The relative rate of leaf area expansion of potato decreased with thermal time, but the reduction was nearly linear up to a leaf area index (L) of 1.0; however, single leaf area of potato increased nearly linearly with thermal time from 5 to 15 m^2kg^{-1} at 50% emergence, from 20 to 25 m^2kg^{-1}at 155°Cd, and then decreased slightly [90]. Potato genotypes with warmer canopies under irrigated conditions are predicted to be less susceptible to drought than genotypes with cooler canopies indicating the rate of transpiration-driven cooling of the leaves.

Canopy photosynthesis (CA): CA on ground cover basis differed significantly among the genotypes, and Pungo had higher values than other genotypes, which ranged between 1.72 and 4.34 g CO_2 m^{-2} $hr.^{-1}$ during 1984 and 1985, respectively [87]. Mean adaxial and abaxial stomatal conductance was 0.86 and 1.46 cm sec^{-1}. Stomatal conductance did not appear to limit gas exchange in potato leaves. Dry matter partitioning to tubers ranged from 8.9% (Pungo) to 55.5% (Atlantic) 67 DAPS and the tuber yield ranged from 9.6 to 27.8 MT/ha, indicating the suitability of cultivar Atlantic for growing in a warm climate. Under stress environment, the leaf area index (LAI) of early cultivars had increased from a value of 3.9 to 5.3 accounting 7% greater cumulative light interception, which consequently enhanced the transpiration rate under moisture stress condition declines light use efficiency and low harvest index in potato. In cv. Dasiree, the dry matter partitioning pattern exhibited earlier and faster tuber filling than the cultivars of same maturity group, where a greater proportion of assimilates diverted toward leaf growth [91].

Root traits: Wishart et al. [92] observed wider variation in rooting traits of a range of potato genotypes including the European tetraploid potato, diploid, and *Neotuberosum* lines The total root length per plant varied from 38 m (Tuberosum variety Pentland Dell) to >100 m (Phureja variety Mayan Twilight). Root thickness and distribution also varied significantly among cultivars with the Phureja line (Mayan Gold) having the longest and thinnest roots (based on ratio of stolon root number with stolon root weight). Number of stolon roots and basal roots among the cultivars (Phureja lines, Mayan Gold) indicated them having significantly more of both these root groups. Further, a significant difference in the relative proportions of basal to stolon root was apparent with Phureja line and Mayan Gold. The largest proportion of basal roots of these genotypes as compared with the other cultivars tested suggested the existence of potential genetic difference in resources partitioning. Hence, improvement in root traits such as root depth and root length density [93] is important for developing cultivars for rain-fed condition. Studying root system in potato is complicated, root-pulling resistance has been proposed as a practical measure to quantify root development. A positive correlation was found between root pulling resistance and tuber yield was noted [94]. Root mass correlates well with leaf mass and tuber yield [95] and justifies that the errectophile, smaller but greater in number of leaves per plant, suits good for rain-fed crop.

Photosynthesis factors: Bhagsari [87] compared the single leaf net photosynthesis determined at 30 ± 2°C by enclosing attached fully mature leaves of sweet potato, cassava, and yam and found the 1.10, 0.70, and 0, 30 mg CO_2 m^{-2} s^{-1}, respectively. The mean canopy photosynthesis rates, expressed on leaf area basis, for sweet potato, cassava, and yam, were 0.18, 0.38, and 0.17 mg CO2 $m^{-2}s - ^{2}$, respectively. With increase in leaf age from 20 to 60 days, the single leaf photosynthetic rate increased in all these crops; however, partitioning of assimilates to root was found higher in cassava. The photosynthetic rate can be improved through breeding [96], and progress could be achieved for high photosynthetic CO_2 exchange rate [97]. Cieply [98] concluded that assimilation rates can be used as a physiological criterion for rapid selection in potato breeding. Mol and Henniger [99] measured rates of "4CO_2 uptake by 18 clones of potato under standard conditions. The stomatal number per leaf at upper surface was found to range from 4 (A6948) to 50 (A66107–51) among the four genotypes evaluated and 130 and 204 at lower surface of the respective genotypes. Similarly, the stomatal apparatus area also showed variation ranging from 0.1 to 2.2 at upper surface and 6.8 and 9.6 at lower surface of respective potato clones. However, the CO_2 uptake was found higher in clone A-6948 (9.2) as compared with A-66107-51 (7.6 mg CO_2mg^{-1} chl h^{-1}), which gives a surprising contribution to total carbon assimilation that cv. Lemhi had an unusually high rate of CO_2 assimilation through the upper leaf surface, and A6948–4 an unusually high rate through the lower leaf surface. Hence, with proper breeding approach, these two traits can be combined for enhancing high carbon assimilation rates of Lemhi's through upper leaf surface and A6948–4's through lower leaf surface.

9. Conclusion

Every crop encounters several production constraints from seed to harvest, which is tactfully manipulated by researcher by developing genotype and suitable production technology for varied environments, which frequently reflect wider yield variations. Because, the genotype by environment interaction reduces the association between the phenotypic and genotypic values and leads to bias in the estimation of gene effects and combining ability for various characters that are sensitive to environmental fluctuations less reliable for selection. In maize, commercial grain yields have improved nearly sixfold and the genetic component of the improvement has been estimated as approximately 60%. The changes in leaf canopy size and architecture account for only a minor portion of the improvement. The majority of the improvement in source capacity is due to visual and functional stay-green. Functional stay-green and the sink establishment dynamics still represent opportunities for yield improvements [100]. The 2.13-fold increase in ERA hybrid grain yield represents a 113% improvement in dry matter accumulation (DMA). The increase in DMA (i.e., the "source") can be attributed, in part, to quantifiable changes in light interception due to increased leaf area index (LAI) and changes in light utilization due to more erect upper leaves. These improvements on the source side were accompanied by improvements in sink establishment dynamics permitted plants to maintain harvest index (HI) of ~50% even when grown under stress conditions.

In potato, a hypothetical attainable yield estimated for different *kharif* growing region of India with existing growing period by changing canopy cover (100%) duration extended for 10–40 days, the GDD accumulation during the growing season could be enhanced up to 800 (Dharwad), 585 (Hassan), 406 (Ooty), 602 (Shimla), and 886 (Srinagar) additionally. Under the condition of harvesting of the additional heat units and converted in terms of dry matter, the attainable yield (25.4, 28.7, 31.7 and 34.7 t/ha, respectively) at Dharwad, Shimla (31.9, 34.9, 38.3. and 42.0 t ha^{-1}) could be obtained. Hence, identification mega environments facilitates the genotypic effect stable and environment effect zero for different production aspects. As physiological breeding coupled with mega environments identification would result in both genotypic and phenotypic effect stable, it can be considered as a sustainable smart farming.

10. Future prospective

In order to exploit yield determinants, the future perspective is to identify mega environments for trait of interest, breed varieties with identified traits, testing of developed varieties specifically for single, multiple, and mega environments based on the traits segregation pattern rather than testing them across environments without targeting the traits of interest in the beginning of breeding program.

Author details

Raja Shankar[1*], Panamanna Mahadevan Govindakrishnan[1], Shashi Rawat[1] and Joseph Sherly[2]

1 ICAR-Central Potato Research Institute, Shimla, India

2 Pandit Jawahar Lal Nehru College of Agriculture and Research Institute, Karaikal, India

*Address all correspondence to: raja.shankar@icar.gov.in

References

[1] Beukema HP, Van der DE Z. Introduction to Potato Production. Vol. 13-23. The Netherland: Pudoc, Wageningen; 1990. pp. 42-52

[2] Bradshaw JE. A genetic perspective on yield plateau in potato. Potato Journal. 2009;**36**(3-4):79-94

[3] Mackay GR. An agenda for future potato research. Potato Research. 1996;**39**:387-394

[4] Haverkort A. Yield levels of potato crops in Central Africa. Agricultural Systems. 1986;**21**:227-235

[5] Haverkort AJ, Harris PM. A model for potato growth and yield under tropical highland conditions. Agricultural & Forest Meteorology. 1987;**39**:271-282

[6] Rawat S, Chauhan RK, Govindakrishnan PM. Potential yield estimation of potato: Comparison of estimates of infocrop potato and versteeg and van keulen methods. Potato Journal. 2007;**34**(1-2):127-128

[7] Gawronska H, Dwelle RB, Pavek JJ, Rowe P. Partitioning of photoassimilates by four potato clones. Crop Science. 1984;**24**:1031-1036

[8] Burton WG. Challenges for stress physiology in potato. American Potato Journal. 1981;**58**:3-14

[9] Kooman PL, Haverkort AJ. Modelling development and growth of the potato crop influenced by temperature and daylength: LINTUL-POTATO. In: Haverkort AJ, MacKerron DKL, editors. Ecology and Modelling of Potato Crops under Conditions Limiting Growth. Dordrecht: Kluwer Academic Publishers; 1995. pp. 41-60

[10] Timlin D, Lutfor Rahman SM, Baker J, Reddy VR, Fleisher D, Quebedeaux B. Whole plant photosynthesis, development and carbon partitioning in potato as a function of temperature. Environment Journal. 2006;**98**:1195-1203

[11] Petrovics DM, Belic M, Banjac B, Boskovic J, Zecevic V, Pejic B. The variation of yield components in wheat (*Triticum aestivum* L.) in response to stressful growing conditions of alkaline soil. Genetika. 2010;**42**(3):545-555

[12] Hassanpanah D, Azimi J. Yield stability analysis of potato cultivars in spring cultivation and after barley harvest cultivation. American-Eurasian Journal of Agricultural & Environmental Sciences. 2010;**9**(2):140-144

[13] Sabaghniaa N, Dehghania H, Sabaghpourb SH. Nonparametric methods for interpreting genotype×environment interaction of lentil genotypes. Crop Science. 2006;**46**:1100-1106

[14] Pushkarnath. Potato in Subtropics. New Delhi: Orient Longman; 1976

[15] Shaner WW, Philipp PF, Schmehl WR. Farming Systems Research and Development Guidelines for Developing Countries. USA: Westview Press, Boulder, Co.; 1982

[16] Cooper M, Hammer GL. Plant Adaptation and Crop Improvement. CAB International: Wallingford, UK; 1996

[17] Yan W, Hant LA, Qinglai S, Szalvincs Z. Cultivar evaluation and mega environment investigation based on the GGE Biplot. Crop Science. 2000;**40**:597-605

[18] Griffith AJF, Miller JH, Suzuki DT, Lewontin RC, Gelbart WM. An Introduction to Genetic Analysis. New York, NY: WH Freeman and Company; 1996

[19] Yan W, Rajcan I. Biplot analysis of test sites and trait relations of soybean in Ontario. Crop Science. 2002;**42**:11-19

[20] Yan W, Kang MS, Ma B, Woods S, Cornelius PL. GGE Biplot vs. AMMI analysis of genotype-by-environment data. Crop Science. 2007;**47**:643-653

[21] Kroonenberg PM. Introduction to Biplots for GxE Tables. Department of Mathematics, Research Report 51. Australia: University of Queensland; 1995

[22] Gauch HG. Statistical Analysis of Regional Yield Trials: AMMI Analysis of Factorial Designs. Amsterdam, The Netherlands: Elsevier; 1992

[23] Ma BL, Yan W, Dwyer LM, Frégeau-Reid J, Voldeng HD, Dion Y, et al. Graphic analysis of genotype, environment, nitrogen fertilizer, and their interaction on spring wheat yield. Agronomy Journal. 2004;**96**:169-180

[24] Ludlow MM, Muchow RC. Critical evaluation of the possibilities for modifying crops for high production per unit of precipitation. Advances in Agronomy. 1990;**43**:107-153

[25] Lopes MS, Reynolds MP. Partitioning of assimilates to deeper roots is associated with cooler canopies and increased yield under drought in wheat. Functional Plant Biology. 2010a;**37**:147-156

[26] Amani I, Fischer RA, Reynolds MP. Canopy temperature depression association with yield of irrigated spring wheat culti vars in a hot climate. Journal of Agronomy & Crop Science. 1996;**176**:119-129

[27] Monteith JL, Unsworth MH. Principles of Environmental Physics. 2nd ed. London: Edward Arnold; 1990

[28] Monteith JL. Climate and efficiency of crop production in Britain. Philosophical Transactions of the Royal Society of London Series B. 1977a;**281**:277-294

[29] Monteith JL. Climate and the efficiency of crop production in Britain. Philosophical Transactions. Royal Society of London. 1977b;**281**:277-297

[30] Sinclair TR, Muchow RC. Radiation use efficiency. Advances in Agronomy. 1999;**65**:125-265

[31] Belanger G, Walsh JR, Richards JE, Milburn PH, Ziadi N. Tuber growth and biomass partitioning of two potato cultivars grown under different N fertilization rates with and without irrigation. American Journal Potato Research. 2001;**78**:109-117

[32] Jefferies RA, MacKerron DKL. Radiation interception and growth of irrigated and droughted potato (*Solanum tuberosum*). Field Crops Research. 1989;**22**:101-102

[33] Vos J. The nitrogen response of potato (*Solanum tuberosum* L.) in the field: Nitrogen uptake and yield, harvest index and nitrogen concentration. Potato Research. 1997;**40**:237-248

[34] Evans LT, Dunstone RL. Some physiological aspects in wheat. Australian Journal of Biological Sciences. 1970;**23**:725-741

[35] López-Castañeda CR, Richards RA, Farquhar GD, Williamson RE. Seed and seedling characteristics contributing to variation in early vigor among temperate cereals. Crop Science. 1996;**36**:1257-1266

[36] Bush MG, Evans LT. Growth and development in tall and dwarf isogenic lines of spring wheat. Field Crops Research. 1988;**18**:243-370

[37] Keyes GJ, Paolillo DJ, Sorrells ME. The effects of dwarfing genes Rht1 and Rht2 on cellular dimensions and rate of leaf elongation in wheat. Annals of Botany. 1989;**64**:683-690

[38] Rawson HM, Gardner PA, Long MJ. Sources of variation in specific leaf area in wheat grown at high temperature. Australian Journal of Plant Physiology. 1987;**14**:287-298

[39] Allard RW. Principles of Plant Breeding. New York: John Wiley & Sons; 1960

[40] Ceccarelli S. Main stages of a plant breeding programme. In: Ceccarelli S, Guimaraes EP, Weltzien E, editors. Plant Breeding and Farmer Participation. Rome: FAO; 2009. pp. 63-74

[41] Gepts P. A comparison between crop domestication, classical plant breeding, and genetic engineering. Crop Science. 2002;**42**:1780-1790

[42] Donald CM. In search of yield. Journal of Australian Institute Agricultural Science. 1962;**28**: 171-178

[43] Donald CM. The breeding of crop ideotype. Euphytica. 1968;**17**:385-403

[44] Fischer RA. Optimising the use of water and nitrogen through breeding of crops. Plant and Soil. 1981;**58**:249-278

[45] Hubick KT, Farquhar GD, Shorter R. Correlation between water-use efficiency and carbon isotope discrimination in diverse peanut (Arachis) germplasm. Australian Journal of Plant Physiology. 1986;**13**:803-816

[46] Jordan WR, Dugas WA, Shouse PJ. Strategies for crop improvement for drought-prone regions. Agricultural Water Management. 1983;**7**:281-299

[47] Shorter R, Lawn RJ, Hammer GL. Improving genotypic adaptation in crops – A role for breeders, physiologists and modelers. Experimental Agriculture. 1991;**27**:155-175

[48] Rasmusson DC. A plant breeder's experience with ideotype breeding. Field Crops Research. 1991;**26**:191-200

[49] Sheeshy JE, Dionora MJA, Mitchell PL. Spikelet numbers, sink size and potential yield in rice. Field Crops Research. 2001;**71**:77-85

[50] Haverkort AJ, Kooman PL. The use of systems analysis and modelling of growth and development in potato ideotyping under conditions affecting yields. Euphytica. 1997;**94**:191-200

[51] Jackson P, Robertson M, Cooper M, Hammer G. The role of physiological understanding in plant breeding; from a breeding perspective. Field Crops Research. 1996;**49**:11-37

[52] Simmonds NW. Introgression and incorporation strategies for the use of crop genetic resources. Biological Reviews. 1993;**68**:539-562

[53] Falconer DS. Introduction to Quantitative Genetics. 3rd ed. Harlow, Essex, UK/New York: Longmans Green/ John Wiley & Sons; 1989

[54] Sojka RE, Slotzy SH, Fischer RA. Seasonal drought response of selected wheat cultivars. Agronomy Journal. 1981;**73**:838-844

[55] Cooper M, DeLacy IH, Eisemann RL. Recent advances in the study of genotype × environment interactions and their

application to plant breeding. In: Imrie BC, Hacker JB, editors. Focused Plant Improvement: Towards Re-Sponsible and Sustainable Agriculture. Proc. 10th Aust. Plant Breeding Conf. Vol. 1. Canberra: Organising Committee. Aust. Convention and Travel Service; 1993. pp. 116-131

[56] Falconer DS. The problem of environment and sclection. The American Naturalist. 1952;**86**:293-298

[57] Morgan JM. Osmoregulation and water stress in higher plants. Annals of Review Plant Physiology. 1984;**35**:299-399

[58] Richard A. Physiological traits used in the breeding of new cultivars for water scarce environments. Agricultural Water Management. 2006;**80**(1-3):197-211

[59] Reynolds MP, Manes Y, Izanloo A, Langridge P. Phenotyping for physiological breeding and gene discovery in wheat. Annals of Applied Biology. 2009;**155**:309-320

[60] Pinto RS, Reynolds MP, Mathews KL, McIntyre CL, Olivares Villegas JJ, Chapman SC. Heat and drought adaptive QTL in a wheat population designed to minimize confounding agronomic effects. Theoretical and Applied Genetics. 2010;**121**:1001-1021

[61] Jovovic Z, Velimirovic A, Dolijanovic Z, Silj M, Zejak D. Possibility of summer planting of potato in agroecological conditions of Podgorica. In: Fifth International Scientific Agricultural Symposium. Agro System. 2014. pp. 433-438

[62] Khurana SMP, Singh RA, Kalay DM. In: Maramorosch K, Raychaudhuri JP, editors. Mycoplasma Diseases of Crops, Basic and Applied Aspects. New York, NY: Springer-Verlag; 1988. p. 285

[63] Bradshaw JE, Mac Kay GR. Breeding strategies for clonally propagated crops. In: Bradshaw JE, MacKay GR, editors. Potato genetics. Wallingford, UK: CAB International; 1994. pp. 467-497

[64] Gopal J. True potato seed: Breeding for hardiness. In: Proceedings of the Sixth Triennial Congress of the African Potato Association, Agadir, Morocco 5-10 April 2004. 2004. pp. 39-57

[65] Hawkes JG. Origins of cultivated potatoes and species relationships. In: Bradshaw JE, Mackay GR, editors. Potato genetics. Wallingford: CAB International; 1994. pp. 3-42

[66] Joseph TA, Birhaman R, Gopal J, Sood SK. Genetic divergence in new potato genotypes. Journal of Indian Potato Association. 2005;**26**:119-125

[67] Raj BT, Kumar D, Minhas JS. Field evaluation of heat tolerant genotypes and commercial varieties against leaf hopper and mite damage on potato. Journal of Indian Potato Association. 2004;**31**:34-37

[68] Patel CK, Patel PT, Chaudhari SM. Effect of physiological age and seed size on seed production of potato in North Gujarat. 2008;**35**(1&2):85-87

[69] Minhas JS, Kumar D, Joseph TS, Raj BT, Paul Khurana SM, Pandey SK, et al. Kufri Surya: A new heat-tolerant potato variety suitable for early planting in north-western plains, peninsular India and processing into French fries and chips. Potato Journal. 2006;**33**(1-2):35-43

[70] Kumar D, Minhas JS, Singh BP. Physiological basis of tolerance to heat stress in advance potato hybrid HT/92-621. In: National Seminar on Plant Physiology. Pune, India: Indian Society of Plant Physiology, New Delhi and Department of Botany, University of Pune; 2004. p. 74

[71] Verma SC. Potato Processing in India. Technical Bulletin. Shimla: Central Potato Research Institute; 1991. pp. 24-34

[72] Smith O. Effect of cultural and environmental conditionson potato processing. In: Talburt WF, Smith O, editors. Potato Processing. New York: Van Nostrand Reinhold Company; 1987. pp. 73-74

[73] Patel NH, Patel RN, Singh SV, Pandey SK, Patel JM, Patel SB. Performance of exotic and Indian potato varieties in Gujarat for processing into french fries. Potato Journal. 2007;**34**(1-2):51-52

[74] Pandey SK, Singh SV, Kumar D, Marwaha RS, Manivel P, Kumar P. Performance of newly released 'Kufri Chipsona 3' Indian potato variety during different crop seasons in west-central plains. Indian Journal of Agricultural Sciences. 2008;**78**(2):116-121

[75] Singh RK, Singh JP, Lal SS. Dry matter partitioning relative to development in high yielding Indian potato cultivars under short day tropical conditions. Potato Journal. 2008;**35**:161-166

[76] Trehan SP, Singh BP. Nutrient efficiency of different crop species and potato varieties – in retrospect and prospect. Potato Journal. 2013;**40**(1):1-21

[77] Barber SA. A diffusion and mass-flow concept of soil nutrient availability. Soil Science. 1962;**93**:39-49

[78] Trehan SP, Sharma RC. External phosphorus requirement of different potato (*Solanum tuberosum*) cultivars resulting from different internal requirements and uptake efficiencies. Indian Journal of Agricultural Science. 2003;**73**(1):54-56

[79] Hoekstra AY, Chapagain AK. Water footprints of nations: Water use by people as a function of their consumption pattern. Water Resources Management. 2007;**21**:35-48

[80] Wolf S, Olesinski AA, Rudich J, Marani A. Effect of high temperature on photosynthesis in potatoes. Annals of Botany. 1990;**65**:179-185

[81] Ozturk G, Yildirim Z. Heritability estimates of some quantitative traits in potatoes. Turkish Journal of Field Crops. 2014;**19**(2):262-267

[82] Dickinson TA, Parker WH, Strauss RE. Another approach to leaf shape comparisons. Taxon. 1987;**36**(1):1-20

[83] Hue SM, Chandran S, Boyce AN. ISHS Acta Horticulturae: Asia Pacific symposium on postharvest research, education and extension variations of leaf and storage roots morphology in *Ipomoea batatas* L. (sweet potato) cultivars. In: ISHS Acta Horticulturae, 2010. Asia Pacific Symposium on Postharvest Research, Education and Extension; 2010

[84] Scoffoni C, Rawls M, McKown A, Cochard H, Sack L. Decline of leaf hydraulic conductance with dehydration: Relationship to leaf size and venation architecture. Plant Physiology. 2011;**156**:832-843

[85] Brodribb TJ, Field TS, Jordan GJ. Leaf maximum photosynthetic rate and venation are linked by hydraulics. Plant Physiology. 2007;**144**:1890-1898

[86] Roth-Nebelsick A, Uhl D, Mosbrugger V, Kerp H. Evolution and function of leaf venation architecture: A review. Annals of Botany. 2001;**87**:553-556

[87] Bhagsari AS. Photosynthesis and stomatal conductance of selected root

crops as related to leaf age. Crop Science. 1988;**28**:902-906

[88] Nunes JCS, Fontes PCR, Araujo EF, Sediyama C. Crescimento da batateira e absorcao de macronutrients influenciadospelossistemas de prepare de solo e irrigacao. Pesquisa Agropecuaria Brasileira. 2006;**41**(12):1787-1792

[89] Jamieson PD, Semenov MA, Brooking IR, Francis GS. Sirius: A mechanistic model of wheat response to environmental variation. European Journal of Agronomy. 1998;**8**:161-179

[90] Van Delden A, Pecio A, Haverkort AJ. Temperature response of early foliar expansion of potato and wheat. Annals of Botany. 2000;**86**:355-369

[91] Spitters CJT, Schapendonk AHCM. Evaluation of breeding strategies for drought tolerance in potato by means of crop growth simulation. Plant and Soil. 1990;**123**:193-203

[92] Wishart JTS, George LK, Brown JA, Thompson G, Ramsay JE, Bradshaw PJ, et al. Variation in rooting habit of potatoes potential for improving resource capture. In: International Symposium Root Research and Applications, Root RAP, 2-4 September 2009. Vienna, Austria: Boku; 2009. pp. 1-4

[93] Ekanayake IJ, Midmore DJ. Genotypic variation for root pulling resistance in potato and its relationship with yield under water-deficit stress. Euphytica. 1992;**61**:43-53

[94] Wall GW, Garcia RL, Kimball BA, Hunsaker DJ, Pinter PJ Jr, Long SP, et al. Interactive effects of elevated carbon dioxide and drought on wheat. Agronomy Journal. 2006;**98**:354-381

[95] Deguchi T, Naya T, Wangchuk P, Itoh E, Matsumoto M, Zheng X. Aboveground characteristics, yield potential and drought tolerance in "Konyu" potato cultivars with large root mass. Potato Research. 2010;**53**:331-340

[96] Crosbie TM, Pearce RB, Mock JJ. Recurrent phenotypic selection for high and low photosynthesis in two maize populations. Crop Science. 1981;**21**:736-740

[97] Mahon JD, Hobbs SL. Selection of peas for photosynthetic CO2 exchange rate under field conditions. Crop Science. 1981;**21**:616-621

[98] Cieply J. The Productivity of Photosynthesis of Several Varieties of Spring Barley and Potatoes as an Index of their Fertility (Abstr.). Krakow, Poland: Rep Acad Agric; 1976

[99] Mol L, Henniger AW. Genotypische Photosynthescratevon Kartoffeln und ihre Mogliche RoUe fur die Ertagsbildung. Photosynthetica. 1978;**12**:51-61

[100] Lee E, Tollenaar M. Physiological basis of successful breeding strategies for maize grain yield. Crop Science. 2007;**47**(3):202-215

Chapter 5

Perspective Chapter: Recent Advances in Nanotechnology, Nanomaterials, Nanofertilizers and Smart Farming

Mohammed Nagib Hasaneen

Abstract

From survey of literature showing a traditional farm practicing which leads to loses in all field crops, it was thought of interest to study novel and new nanotechnologies in farming of field crops to increase yield quantity and quality, to reduce the use of chemical fertilizers, to reduce water irrigation, to exclude the use of pesticides for control of plant diseases, to produce a biosafety crops and finally to produce a safe crops. All these can easily take place by introducing smart farming, the nanofertilizer and nanodrug delivery systems for treatment and control of all plant diseases.

Keywords: antibiotics, carbon nanotubes, chitosan nanoparticles, field crops, Nanofertilizers, Nanodrug delivery strategies, smart farming, plant diseases

1. Introduction

1.1 NPK-nanofertilizers

Nanotechnology has evolved over the last few decades to occupy everyday life, and agriculture is one of the areas where nano-applications have recently reached.

The massive increase in human populations all over the globe means that we need to provide more food from the same area of cultivated lands. This means that we need to produce better crops and increase supplies with the same resources present. For this reason, new methods to increase crop productivity and lower fertilizer consumption are now being researched.

Artificial fertilizers are identified as inorganic fertilizers which are formed in appropriate concentrations to supply three chief elements: nitrogen, phosphorus and potassium (N, P and K) for different crops and growing conditions. NPK-inorganic fertilizers are vital for plant growth and development. N (nitrogen) stimulates leaf growth and is found in proteins and chlorophyll, P (phosphorus) improves root, flower and fruit development, and K (potassium) enhances stem and root growth and the production of proteins. However, plants utilized only about 30–60%, 10–20% and 30–50% the applied dose of N, P and K, respectively, and the rest is lost to the environment causing serious contamination to soil and water as well as substantial economic and resource losses. To minimize these conventional fertilizer demerits and utilize the major proportion of the applied dose of the chemical, nanotechnology can be applied by encapsulating the nutrients in nanomaterials, coated with a thin protective film or delivered as emulsions or nanoparticles.

1.2 What is a nanofertilizer?

Nanofertilizers are nutrient carriers in the dimension of 1–100 nm. "Nano" refers to one-billionth of a meter or one-millionth of a millimeter. When the size gets reduced, the surface area has tremendously increased. Nanofertilizers are a nano-structured formulation of fertilizers that release nutrients into the soil gradually and in a controlled way. The nutrient uptake efficiency can be increased by using nano-based slow-release or controlled release fertilizers which can lead to significantly reduce the wastage of nutrients. The nanomaterials may be applied either in the soil nutrition or by foliar application by developing the formulations, i.e., coated, encapsulated or buried in the nanomaterials.

1.3 NPK-fertilizers coated or encapsulated with nanoparticles

In our research work, we used chitosan nanoparticles loaded with NPK as foliar fertilizer for wheat plants. We used three concentrations of the nanofertilizer which are 10, 25 and 100%. During foliar application, all pots were covered to prevent the entry of nanofertilizers to the soil. The results showed that nanofertilizers induced significant increases in all growth and yield variables determined when compared with the control (water) or normal-fertilized wheat plants. To our surprise, nanofertilizers decreased the life span of the crop from 170 days for control and normal fertilized plants to just 130 days (a decrease of 23.5%). In addition, these results enabled wheat plants to grow in pure sandy soil with efficient crop productivity. When we studied the uptake of the nanoparticles by the plant through transmission electron microscopy, nanoparticles were found to accumulate in sieve tubes of phloem tissue, while xylem vessels appeared with zero nanoparticles. The lowest concentration (10%) produced the best results as a nanofertilizer for wheat plants. Foliar application of nanofertilizers showed a significant increase in total saccharide content of wheat grains. The magnitude of increase was most pronounced in the nanofertilized wheat plants grains, particularly at 10% nanofertilizer than in normal fertilized ones. Significant decrease in each protein and nitrogen content of the wheat grains was induced when wheat plants were with increasing levels of either normal or nanofertilizer as compared with the control ones. On the other hand, the element content of wheat grains especially potassium and phosphorus contents was significantly increased in nanofertilized wheat plants.

1.4 Foliar application

In another trial, we used carbon nanotubes loaded with NPK and compared them with chitosan nanoparticles loaded with NPK and used both of these types as fertilizers for French bean. In this trial, we tried three different application methods of the nanofertilizers used: soil incorporation, seed priming and foliar application. For soil incorporation, nanoparticles were mixed with the soil. For seed priming, the seeds were soaked in nanosolutions for 30 minutes prior to planting. For foliar application, nanofertilizers were foliary sprayed on the sixteenth day after planting.

The results showed that foliar application gave the best results for growth and yield of the plants. Also, foliar application of both nanofertilizers reduced the life span of the plant to 80 days when compared with 110 days for soil and seed priming treatments. For uptake and translocation studies, chitosan nanoparticles appeared in the phloem tissue only and were absent from the xylem vessels. However, carbon nanotubes appeared in both xylem and phloem tissues. Foliar application of nanofertilizers resulted in progressive significant increases in total carbohydrate, protein and vitamin C contents of the yielded French bean seeds, when compared with the control seeds and with the seeds of French bean plants treated by seed priming and soil incorporation. The best nanofertilizer in this trial appeared to be chitosan nanoparticles loaded with NPK (10%), compared with carbon nanotubes NPK (20 μg/L) (**Figure 1**).

1.5 Pros and cons

NPK-nanofertilizers promise to be a revolution in the fertilizer industry. The high efficiency of crop production, better seed quality, better yield attributes and productivity are key elements when we consider the application of NPK-nanofertilizers. But, until now, few studies have dealt with the possible phytotoxic effects of nanofertilizers to plants. The major threat here is that plants, especially food crops, enter the food chain and the bioaccumulation of nanoparticles may reach animals or humans or may reside in the environment. Possible ways to study the phytotoxic effects of nanofertilizers are now under

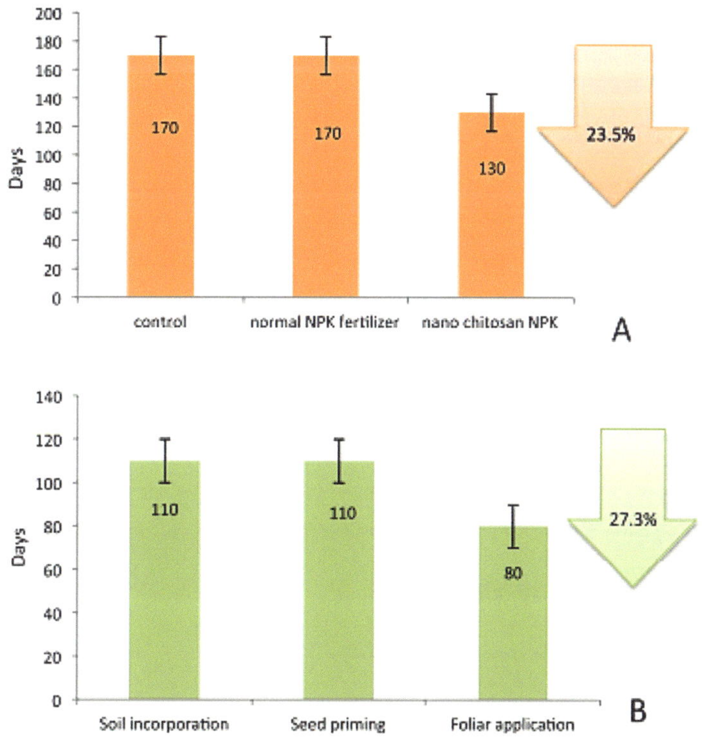

Figure 1.
A: Effects of normal NPK fertilizer and nanoengineered chitosan NPK fertilizer on the life span of wheat plants grown in sandy soil. B: Effects of different methods of application of NPK nanofertilizers on the life span of French bean plants grown in clay-sandy soil.

Parameter / Treatment	Soil incorporation			Seed priming			Foliar application		
	Total carbo-hydrates	Total protein	Vitamin C	Total carbo-hydrates	Total protein	Vitamin C	Total carbo-hydrates	Total protein	Vitamin C
Control	159.88	14.00	6.07	159.88	14.00	6.07	159.88	14.00	6.07
Nano Chitosan NPK (10%)	Dead (no yield)			150.80	13.15	5.50	227.34	22.80	10.00
Carbon nanotubes NPK (20µg/L)	152.50	13.38	5.72	155.60	13.85	6.00	220.10	20.00	8.38

Figure 2.
Effects of different methods of application of NPK nanofertilizers on total carbohydrates (mg glucose equivalent/g dry weight), total protein (mg/g dry weight) and vitamin C (mmole/g dry weight) contents of French bean yielded seeds. (Data from experiments in 2016 in the plant physiology laboratory, Faculty of Science, Mansoura University, Egypt).

research. Also, the safety of long-term nanofertilizer consumption is yet to be confirmed. Possible measures and safety levels must be defined for each nanofertilizer used.

1.6 Current research

Up-to-date studies have found that either negative, insignificant or positive effects of nanoparticles on plant may depend on the type of nanoparticles, plant and soil. Depending on their physical and chemical properties, nanoparticle bioaccumulation in plants is specific. While some studies report beneficial effects on some plant species, the overall negative effect of the accumulation of these nanoparticles in the soil and plants might exceed the minor beneficial temporary effects. The main negative effects may involve the inhibition of growth, oxidative stress and genetic alteration due to the interactions between the plants and nanoparticles. For plant morphology, nanoparticles were found to alter morphological features of plants in vital organs such as the roots and leaves in addition to their effect on seed germination. Many nanoparticles can be translocated within plants and enter the food chain, be available for trophic transfer and become available in food for humans and animals. Many nanoparticles are shown to be toxic to humans, and uptake of nanoparticles in plants poses major safety concerns. Nanomaterials can become an environmental pollutant that might be conducive to irreversible or undesirable modifications with potentially harmful consequences on plants, animals and humans alike (**Figure 2**).

2. Experimental methodologies

2.1 Preparation of nanomaterials

2.1.1 Preparation of chitosan nanoparticles (Cs)

Chitosan nanoparticles (CS-nanoparticles or CS-PMAA nanoparticles) were prepared according to DeMoura et al. [1] and Hasaneen et al. [2] method by polymerization of methacrylic acid (MAA) in chitosan (CS) solution.

For 12 hours, about 0.2 g of chitosan powder was dissolved in a (0.5 v/v) methacrylic acid aqueous solution under magnetic stirring. Then, about 5 mg of $K_2S_2O_8$

was added to the solution with continued stirring till the solution became clear. After that, the mixture was heated up at 70°C for one hour under magnetic stirring to ensure the formation of chitosan nanoparticles. Finally, to stop the reaction, the solution was cooled in an ice bath.

2.1.2 Preparation of carbon nanotubes (CNTs)

In the present study, CNTs were prepared by using the method of Lee and Seo [3]. To a mixture of sulfuric acid and nitric acid (2:1 v/v), five grams of graphite powder was added slowly and stirred for 30 minutes, then cooled at 4°C. After that, about 25 g of potassium chlorate was slowly added to the solution and stirred for 30 minutes, and then, the mixture was heated for 24 hours at 70°C. The mixture then was left for 3 days, and the floating solution was collected and rinsed with distilled water to 1000 cm^3, stirred for 1 hour and filtered, and finally, the sample was dried.

2.1.3 Preparation of solid lipid nanoparticles (SLN)

Solid lipid nanoparticles were prepared *via* hot homogenization method at a temperature above the melting point of lipid using the solid lipid, Tween 80 as the hydrophilic surfactant and soya lecithin as the lipophilic surfactant [4].

2.1.4 Loading of fertilizers NPK on sequences of nanoparticles

According to Corradini et al. [2, 5], the loading of CS-nanoparticles surface with NPK fertilizers was obtained by dissolving about 0.1 g of urea, 0.02 g of calcium phosphate and 0.06 g of potassium chloride as sources of N, P and K fertilizers into 100 cm^3 of CS-nanoparticle solution for 6 hours under magnetic stirring at room temperature. The pH of resulting solution was between 4.2 and 4.7. Meanwhile, the loading carbon nanotubes surface with NPK fertilizers was achieved by adding about 0.4 g of N, 0.1 g of P and 0.4 g of K into 100 cm^3 of CNTs solution and stirring at 25°C for 6 hours.

2.1.5 Impact of engineered nanomaterials either alone or loaded with NPK on growth and productivity of plants

Abdel-Aziz et al. [6] reported that nanotechnology has become a solution to several problems facing humans nowadays. In agriculture, nanofertilizers play a vital role to minimize environmental pollution problems and enrich crop productivity. In this paper, we study the possible effects of using two different nanoengineered materials chitosan nanoparticles (nano-Cs) and carbon nanotubes (CNTs) either alone or loaded with NPK as fertilizers on French bean plants using two different methods of application, namely foliar application and seed priming. It is apparent from the obtained results that foliar application is the better method for application than seed priming. This is obvious from the improvement of growth, yield, antioxidant system and biochemical content of the yielded seeds of foliary applied plants than in plants treated with seed priming technique as compared with control ones. In addition, foliar treatment shortened the days to harvest without reducing yield by 37.5% (80 days) as compared with control and seed priming treatment (110 days). Of interest, nanochitosan either alone or loaded with NPK improved the growth and yield of the foliary treated plants more than CNTs. In conclusion, nanofertilizers foliar application holds tremendous potential to improve crop productivity.

The key focus areas for nanotechnology agricultural research are drug delivery, nano-biofarming, nanopesticides and nanoherbicides and controlled release of nanofertilizers [7]. Nanofertilizers are nutrient carriers on which nutrient ions can be able to be loaded due to their high surface area and able to release these ions in a minimal dose to the soil according to the needs of the cultivated plant which can reduce environmental pollution problems related with excess release of ions which is apparent such in the case of incorporation of essential elements (NPK) on the surface of chitosan nanoparticles as described by Golbashy et al. [8] and AbdelAziz et al. [9]. There are slow-release and super sorbent nitrogenous and phosphatic fertilizers [10, 11].

Nanofertilizers increase the absorption capacity of plant roots which leads to increased photosynthesis and improved crop production [12]. Benzon et al. [13] showed that the application of nanofertilizer to rice plants led to an increment in both total phenolics content and antioxidants which increased plant nutrition and enhanced crop productivity.

Chitosan nanoparticles (nanochitosan or Cs nanoparticles) are engineered nanomaterials produced from chitosan (a linear hydrophilic polysaccharide) which has been used as a functional biopolymer in pharmaceutics and food [14]. Due to physicochemical properties such as size, surface area and cationic nature, the new form of Cs-nanoparticles provides a variety of possible biological activities [15]. Furthermore, Cs and its derivatives have the ability to stimulate both physiological and biochemical activities in plants from single cells and tissues to molecular level changes at gene expression [16], also play a main role in seed germination and plant growth enhancement [17], increase nutrient uptake by plants [18], increase the content of chlorophyll and chloroplast development which enhances photosynthetic activity and increase crop productivity [19, 20].

Carbon nanotubes (CNTs) are defined as cylinders of carbon atoms at nanoscale diameters and microscale lengths [21] and may be categorized to either single-walled CNTs (SWCNTs) or multi-walled CNTs (MWCNTs) depending on the carbon shells number present in the nanotube [22]. The nanoscale of CNTs provide them with unique properties different from carbon and graphite such as physicochemical characteristics like diameter, length and functionality which gives them varied chemical reactivity [23, 24]. Mondal et al. [25] reported positive effects of low CNTs doses in seed germination, water transport and root development in mustard plants, but negative effects were recorded at high concentrations due to the production of reactive oxygen species (ROS) in cells which caused membrane and cell injury which might lead to cell death [26].

Zheng et al. [27] and Klaine et al. [28] reported that nano-TiO_2 at low doses were able to increase nitrogen metabolism and improve photosynthesis which improves plant growth of spinach plants. Farnia and Ghorbani [29] reported that foliary application of K-nanofertilizer to red bean plants increased the 1000 grain weight, number of grains per pod, biomass yield and grain yield as compared to control ones.

In seed priming technique, seeds are soaked in a specific solution for a period of time (no radicle emergence or breaks in seed coat) and then are used for germination [30]. Priming the seeds with nanoparticles solutions can produce highly resistant seeds and improve its germination and consequently improve seedling growth especially under stressful conditions [31]. Zayed et al. [32] reported that priming of bean seeds with (0.1%, 0.2% and 0.3%) nanochitosan for three hours and germinated under salt stress (100 mM NaCl) enhanced seed germination and radicle length. Also, the content of both proline and chlorophyll a as well as antioxidant enzyme activities of bean seedlings treated with 0.1% nanochitosan showed significant improvement as compared with salt-stressed untreated bean seedlings.

2.1.6 Carbon nanotubes NPK and chitosan nanoparticles NPK fertilizer on productivity of plants

Froggett [33] and Hasaneen et al. [34] stated that nanotechnology has proved its place in agriculture and related industries and the using of nanomaterial has the potential to revolutionize the agriculture with novel tools to enhancing the plant ability to absorb nutrients. The ability of plants to uptake nanomaterials has shown a very recent field of nanoagriculture. Several studies reported that nanomaterials were able to penetrate the plant cell by endocytosis. As mentioned above, we designed this work to investigate the effects of different concentrations of two different types of engineered nanofertilizer specifically carbon nanotubes (CNTs) and chitosan nanoparticles (nano-Cs) coated with NPK on the different growth criteria French bean plants. Herein, the results of the both morphological and anatomical analysis indicate that after about 30 days from the date of planting our experimental conditions, nanofertilizers either alone or in combination with conventional fertilizers significantly improved the growth and biomass of plant compared to unfertilized plants. Transmission electron microscopy images (TEM) of the plant leaves indicated the presence of engineered nanomaterials in vascular bundles specifically in sieve tubes of phloem elements in case of nanochitosan-NPK and in both xylem vessels and sieve tubes in case of CNTs- NPK. Overall, after investigation, we conclude that low nanofertilizers doses have seen to be beneficial, improving water absorption and nutrients uptake, found to enhance the plant growth.

Chitosan is a linear hydrophilic polysaccharide that is biocompatible, biodegradable, non-toxic nature biopolymer and reacts with bioactive molecules [14, 35, 36]. Chitosan nanoparticles with a size about 78 nm can be used for controlled release of NPK fertilizer sources such as urea, calcium phosphate and potassium chloride [5]. Carbon nanotubes have an important position due to their unique physicochemical properties such as length, diameter, atomic configuration, defects, impurities and functionality, which allow them to have wide-ranging conductivity, flexibility, tension strength and chemical reactivity properties [23]. CNTs have positive effects on seed germination, root development and water transport within the plant or no evidence of phytotoxicity when plants treated with low doses. On the other hand, the negative effects can be produced at high concentrations due to the generation of harmful reactive oxygen species (ROS) [25]. The uptake of nanofertilizers into plant cell can occur *via* various ways such as binding to carrier proteins through aquaporin, ion channels or endocytosis [37], and also, they may diffuse apoplectically in the space between the cell wall and plasma membrane and can merge into the simplest then penetrate into vascular system [38]. Engineered nanofertilizers can be transported through the phloem elements when applied foliary. For foliar uptake of nanoparticles, there are two possible pathways, namely cuticular and stomatal pathways [39]. In cuticular pathway, nanoparticles with sizes below 5 nm can be uptaken due to extremely small sizes of cuticular pores [40]. While, in stomatal pathways, larger nanoparticles can be penetrated since the typical stomatal size is in micrometer size range [39].

2.1.7 Nano-drug delivery systems and plant diseases

Hasaneen et al. [41] stated that due to the development of antibiotic resistant strains in pathogenic microbes, there is an increasing in microbial diseases yearly which represents a great challenge to the public health, and this considered as an alarming issue worldwide [42]. It is well known that the current medicinal regime delivers drugs to the site of action or inflammation with unavoidable side effects [43]. With nanodrug delivery system, antimicrobial compounds can be accessed to the targeting site of the microbial pathogens [41]. We concluded from our study that antimicrobial compounds extracted from the local isolate namely *Streptomyces rimosus* and loaded on either chitosan nanoparticles or calcium phosphate nanoparticles were found to be facilitate drug delivery to some bacterial species (*Escherichia coli* ATCC25922, *Staphylococcus aureus* ATCC25923 and *Bacillus cereus* ATCC66331 and the yeast *Candida albicans* ATCC10231. The best incubation period for the production of antifungal and antibacterial compounds, pH, temperature, carbon sources and nitrogen sources was around the third day, 70, 30°C, starch and potassium nitrate, respectively. Gas chromatography–mass spectrometry (GC–MS) analysis was used for the identification of the extracted compounds which revealed to the identification of nine antimicrobial organic acids. The prepared NPs were characterized using transmission electron microscopy (TEM) and Zeta potential analyzer. The tested strains were resistant to solo nanoparticles, but the extracted antimicrobial compounds, especially CaP-NPs, improved the isolated antimicrobial compounds potency causing differential antimicrobial activity. The activity of nanoparticles loaded with antimicrobial compounds was more obvious against bacteria than fungi, against, Gram-positive than Gram-negative bacteria and against *B. cereus* than *S. aureus.*

Nanotechnology offers an effective mean of sustained drug delivery and release with avoidance of the problems of the current delivery systems. For proper manipulation, drugs must have a size such that they can be injected without blocking needles and capillaries [44]. This can be achieved by using either nano-liposomes, nano-gels or micelles. By the aid of nanotechnology, drugs can be either loaded on the surface of nanoparticles or encapsulated and cared within then to the drug destination. By this way, the effective dose of the drug can be lowered several orders of magnitude, which led to the minimization of the drug side effects [45].

Chitosan (CS) itself and its derivatives or in the form of nanoparticles have attracted great attention due to their antifungal and antibacterial activity [46, 47]. The safe CS can interact with polyanions to form complexes and gels [48, 49]. While, CaP are the most important component of human teeth, bone and the biological hard tissues in the form of carbonated hydroxyapatite, which afford stability, hardness and proper function [50, 51]. CaP-NPs were manipulated as successful adjuvant with DNA vaccines [52]. New and aggressive antibiotic-resistant bacteria and parasites call for the development of new therapeutic strategies to overcome the inefficiency of conventional antibiotics and to bypass treatment imitations related to these pathologies. Therefore, the present work focuses on the development and combination of CSNPs and CaPNPs with potent antimicrobial compounds that can aid in delivery of antibiotics to the target sites of drug-resistant microorganisms.

3. Conclusion

In smart farming, summarizing the obtained results, using nanochitosan and carbon nanotubes either singly or loaded with NPK as nanofertilizers, throughout the entire period of the cultivation led to:

a. The best technique used for application of nanofertilizers to plant was foliar application [53, 54].

b. The best nanofertilizer used to field crops was nanochitosan-NPK followed by CNTs-NPK.

c. Giving a percent improvement in the quantity and quality of the yielded crops seeds treated foliary with Cs-NPK and CNTs-NPK was 82% and 84%, respectively.

d. A novel technique, treatment and control of plant disease by nanodrug delivery strategies showed a high percent of recovery from disease with 100% in case of using solid lipid nanoparticles loaded with antibiotic and 80% recovery from disease in case of using chitosan nanoparticles loaded with antibiotic and finally 60% recovery from disease in case of using carbon nanotubes loaded with antibiotic [41, 55].

Author details

Mohammed Nagib Hasaneen
Botany Department, Mansoura University, Mansoura, Egypt

*Address all correspondence to: mohammednagib@mans.edu.eg

References

[1] DeMoura MR, Aouada FA, Mattoso LHC. Preparation of chitosan nanoparticles using methacrylic acid. Journal of Colloid and Interface Science. 2008;**321**:477-483

[2] Hasaneen MNA, Abdel-Aziz HMM, El-Bialy DMA, Omer AM. Preparation of chitosan nanoparticles for loading with NPK fertilizer. African Journal of Biotechnology. 2014;**13**:3158-3164

[3] Lee DW, Seo JW. Preparation of carbon nanotubes from graphite powder at room temperature. 2010. arXiv preprint arXiv:1007.1062

[4] Gazi AS, Krishnasailaja A. Preparation and evaluation of paracetamol solid lipid nanoparticle by hot homogenization method. Nanomedicine Research Journal. 2018;**7**:152-154

[5] Corradini E, DeMoura MR, Mattoso LDC. A preliminary study of the incorporation of NPK fertilizer into chistosan nanaparticles. Express Polymer Letters. 2010;**4**:509-515

[6] Abdel-Aziz HMM, Hasaneen MNA, Omer AM. Impact of engineered nanomaterials either alone or loaded with NPK on growth and productivity of French bean plants: Seed priming vs foliar application. South African Journal of Botany. 2019;**125**:102-108

[7] Agrawal S, Rathore P. Nanotechnology pros and cons to agriculture: A review. International Journal of Current Microbiology and Applied Sciences. 2014;**3**:43-55

[8] Golbashy M, Sabahi H, Allahdadi I, Nazokdast H, Hosseini M. Synthesis of highly intercalated urea-clay nanocomposite via domestic montmorillonite as eco-friendly slow-release fertilizer. Archives of Agronomy and Soil Science. 2016;**63**:84-95

[9] Abdel-Aziz HMM, Hasaneen MNA, Omer AM. Nano chitosanNPK fertilizer enhances the growth and productivity of wheat plants grown in sandy soil. Spanish Journal of Agricultural Research. 2016;**14**:17

[10] Hasaneen MNA, Tourky SMN, Abo-Elwafa GS, Hassan MN. Can cyanobacteria biofertilizer enhance the growth of *Phaseolus vulgaris* plants? Journal of Plant Production. 2015;**2015**:6

[11] Lal R. Promise and limitations of soils to minimize climate change. Journal of Soil and Water Conservation. 2008;**63**:113-118

[12] INIC. Iran Nanotechnology Initiative Council. First nano-organic iron chelated fertilizer invented in Iran. 2014. Available from: http://www.iranreview.org/content/Documents/Iranians_Researchers_Produce_Nano_Organic_Fertilizer.htm

[13] Benzon HRL, Rubenecia MRU, Ultra VU Jr, Lee SC. Nano-fertilizer affects the growth, development, and chemical properties of rice. International Journal of Agronomy and Agricultural Research. 2015;**7**:105-117

[14] Cho Y, Shi R, Borgens BR. Chitosan produces potent neuroprotection and physiological recovery following traumatic spinal cord injury. Journal of Experimental Biology. 2010;**213**:1513-1520

[15] ChandraS,ChakrabortyN,DasguptaA, Sarkar J, Panda K, Acharya K. Chitosan nanoparticles: A positive modulator

of innate immune responses in plants. Scientific Reports. 2015;**5**:15195. DOI: 10.1038/srep15195

[16] Van NS, Minh DH, Anh ND. Study on chitosan nanoparticles on biophysical characteristics and growth of Robusta coffee in green house. Biocatalysis and Agricultural Biotechnology. 2013;**2**:289-294

[17] Nge KL, Nwe N, Steven W. Chitosan as a growth stimulator in orchid tissue culture. Plant Science. 2006;**170**:1185-1190

[18] Dzung NA, Khanh VTP, Dung TT. Research on impact of chitosan oligomer on biophysical characteristics, growth, development and drought resistance of coffee. Carbohydrate Polymers. 2011;**84**:751-755

[19] Dzung NA. Enhancing crop production with chitosan and its derivatives. In Book: Chitin, Chitosan, Oligosaccharides and their Derivatives: Biological Activity and Application. CRC Press, Taylor & Francis; 2010. pp. 619-632

[20] Limpanavech P, Chaiyasuta S, Vongpromek R, Pichyangkura R, Khunwasi C, Chadchawan S, et al. Chitosan effects on floral production, gene expression, and anatomical changes in the *Dendrobium orchid*. Scientia Horticulturae. 2008;**116**:65-72

[21] Petersen EJ, Zhang L, Mattison NT, O'Carroll DM, Whelton AJ, Uddin N, et al. Potential release pathways, environmental fate, and ecological risks of carbon nanotubes. Environmental Science and Technology. 2011;**45**:9837-9856

[22] Popov VN. Carbon nanotubes: Properties and application. Materials Science and Engineering. 2004;**43**:61-102

[23] Jackson P, Jacobsen NR, Baun A, Birkedal R, Kühnel D, Jensen KA, et al. Bioaccumulation and ecotoxicity of carbon nanotubes. Chemistry Central Journal. 2013;**7**:154-175

[24] Philip GC, Keith B, Masa I, Zettl A. Extreme oxygen sensitivity of electronic properties of carbon nanotubes. Science. 2000;**287**:1801-1804

[25] Mondal A, Basu R, Das S, Nandy P. Beneficial role of carbon nanotubes on mustard plant growth: An agricultural prospect. Journal of Nanoparticle Research. 2013;**13**:4519-4528

[26] Moore MN. Do nanoparticles present ecotoxicological risks for the health of the aquatic environment? Environment International. 2006;**32**:967-976

[27] Zheng L, Hong F, Lu S, Liu C. Effect of nano-TiO_2 on strength of naturally aged seeds and growth of spinach. Biological Trace Element Research. 2005;**104**:83-91

[28] Klaine SJ, Alvarez PJJ, Batley GE, Fernandes TF, Handy RD, Lyon DY, et al. Nanomaterials in the environment: Behavior, fate, bioavailability, and effects. Environmental Toxicology and Chemistry. 2008;**27**:1825-1851

[29] Farnia A, Ghorbani A. Effect of K nano-fertilizer and N bio-fertilizer on yield and yield components of red bean (*Phaseolus vulgaris* L.). International Journal of Biosciences. 2014;**5**:296-303

[30] Paparella S, Araújo SS, Rossi G, Wijayasinghe M, Carbonera D, Balestrazzi A. Seed priming: State of the art and new perspectives. Plant Cell Reports. 2015;**34**:1281-1293

[31] Maroufi K, Farahani HA, Moradi O. Evaluation of nano priming on germination percentage in

green gram (*Vigna radiata* L.). Advances in Environmental Biology. 2011;**5**:3659-3664

[32] Zayed MM, Elkafafi SH, Zedan AMG, Dawoud SFM. Effect of nano chitosan on growth, physiological and biochemical parameters of Phaseolus vulgaris under salt stress. Journal of Plant Production. 2017;**8**:577-585

[33] Froggett S. Nanotechnology and agricultural trade on application of nanometer biotechnology on the yield and quality of winter wheat. Journal of Anhui Agricultural Sciences. 2009;**36**:15575-15580

[34] Hasaneen MNA, Abdel-Aziz HMM, Omer AM. Effect of foliar application of engineered nanomaterials: Carbon nanotubes NPK and chitosan nanoparticles NPK fertilizer on the growth of French bean plant. Biochemistry and Biotechnology Research. 2016;**4**:68-76

[35] Dai M, Zheng X, Xu X, Kong YX, Guo G, Luo F, et al. Chitosan alginate sponge: Preparation and application in curcumin delivery for dermal wound healing in rat. Journal of Biomedicine and Biotechnology. 2009;**2009**:595126

[36] Hanafi N. Role of chitosan nanoparticles in targeting Ehrlich tumor cells transplanted in albino mice. International Journal of Biological Sciences. 2012;**2**:6-17

[37] Nair R, Varghese SH, Nair BG, Maekawa T, Yoshida Y, Kumar DS. Nanoparticulate material delivery to plants. Plant Science. 2010;**179**:154-163

[38] Larue C, Pinault M, Czarny B. Quantitative evaluation of multiwalled carbon nanotube uptake in wheat and rape seed. Journal of Hazardous Materials. 2012;**228**:155-163

[39] Eichert T, Goldbach HE. Equivalent pore radii of hydrophilic foliar uptake routes in stomatous and astomatous leaf surfaces–further evidence for a stomatal pathway. Physiologia Plantarum. 2008;**132**:491-502

[40] Eichert T, Kurtz A, Steiner U, Goldbach HE. Size exclusion limits and lateral heterogeneity of the stomatal foliar uptake pathway for aqueous solutes and water-suspended nanoparticles. Physiologia Plantarum. 2008;**134**:151-160

[41] Hasaneen MNA, Abou-Dobara MI, Nabih S, Mousa M. Preparation, optimization, characterization and antimicrobial activity of chitosan and calcium nanoparticles loaded with *Streptomyces rimosus* extracted compounds as drug delivery systems. Journal of Microbiology, Biotechnology and Food Sciences. 2022:e5020-e5020

[42] Jones KE, Nikkita GP, Marc AL, Adam S, Deborah B, John LG, et al. Global trends in emerging infectious diseases. Nature. 2008;**451**(7181):990-993

[43] Chadha S. Nanotechnology and its application. International Journal of Agriculture and Food Science Technology, Nanotechnology and Its Application. 2013;**4**:1011-1018

[44] Hughes GA. Nanostructure-mediated drug delivery. Nanomedicine: Nanotechnology, Biology and Medicine. 2005;**1**:22-30

[45] Campagnolo L, Massimiani M, Palmieri G, Bernardini R, Sacchetti C, Bergamaschi A, et al. Biodistribution and toxicity of pegylated single wall carbon nanotubes in pregnant mice. Particle and Fibre Toxicology. 2013;**10**:1-13

[46] Baka ZA, Abou-Dobara MI, El-Sayed AK, El-Zahed MM. Synthesis, characterization and antimicrobial activity of chitosan/Ag nanocomposite using *Escherichia coli* D8. Scientific Journal for Damietta Faculty of Science. 2020;**9**:1-6

[47] Tsai GJ, Su WH. Antibacterial activity of shrimp chitosan against *Escherichia coli*. Journal of Food Protection. 1999;**3**:239-243

[48] Agnihotri SA, Mallikarjuna NN, Aminabhavi TM. Recent advances on chitosan-based micro-and nanoparticles in drug delivery. Journal of Controlled Release. 2004;**100**:5-28

[49] Kim SK, Rajapakse N. Enzymatic production and biological activities of chitosan oligosaccharides (COS): A review. Carbohydrate Polymers. 2005;**62**:357-368

[50] Dorozhkin SV, Epple M. Biological and medical significance of calcium phosphates. Angewandte Chemie International Edition. 2002;**41**:3130-3146

[51] Epple M, Baeuerlein E. Biomineralisation: Medical and Clinical Aspects. Weinheim: Wiley-VCH; 2007. p. 13

[52] Joyappa DH, Kumar CA, Banumathi N, Reddy GR, Suryanarayana VV. Calcium phosphate nanoparticle prepared with foot and mouth disease virus P1-3CD gene construct protects mice and guinea pigs against the challenge virus. Veterinary Microbiology. 2009;**139**:58-66

[53] Abdel-Aziz HMM, Hasaneen MNA, Omar AM. Effect of foliar application of nano chitosan NPK fertilizer on the chemical composition of wheat grains. Egyptian Journal of Botany. 2018;**58**:87-95

[54] Omar AM. Comparative Studies on the Effects of Chitosan Nanoparticles and Carbon Nanotubes Loaded with NPK on the Growth and Productivity of French Bean. Mansoura, Egypt: Mansoura University; 2017

[55] Helal SH, Hasaneen MNA, El-Zayat MM, Abdel-Aziz HMM. *In vitro* efficacy of antifungal activities either singly or loaded on nanomaterials, as new and novel drug delivery systems on the growth of *Alternaria alternata* and *Botrytis fabae* fungal pathogens. Mansoura. Journal of Biology. 2022;**2022**:41